Gravity

BOOKS OF RELATED INTEREST

Rutley's elements of mineralogy, 13th edn
H. H. Read

Petrology of the igneous rocks, 26th edn
F. H. Hatch, A. K. Wells & M. K. Wells

Metamorphism and metamorphic belts
A. Miyashiro

Petrology of the metamorphic rocks
R. Mason

The interpretation of igneous rocks
K. G. Cox, J. D. Bell & R. J. Pankhurst

Atmospheric processes
J. D. Hanwell

The inaccessible Earth
G. C. Brown & A. E. Mussett

Tectonic processes
D. Weyman

Sedimentology: process and product
M. R. Leeder

Introduction to small-scale geological structures
G. Wilson

Metamorphic geology
C. Gillen

The poetry of geology
R. M. Hazen

Komatiites
edited by N. T. Arndt & E. G. Nisbet

Boundary element methods in solid mechanics
S. L. Crouch & A. M. Starfield

The boundary integral equation method for porous media flow
J. L. Liggett & P. L-F. Liu

Gravity

Chuji Tsuboi

Professor Emeritus, University of Tokyo

London
GEORGE ALLEN & UNWIN
Boston Sydney

Juryoku by Chuji Tsuboi

Translated by Chuji Tsuboi
Originally published in Japanese by Iwanami Shoten, Publishers, Tokyo, 1979

George Allen & Unwin (Publishers) Ltd,
40 Museum Street, London WC1A 1LU, UK

George Allen & Unwin (Publishers) Ltd,
Park Lane, Hemel Hempstead, Herts HP2 4TE, UK

Allen & Unwin Inc.,
9 Winchester Terrace, Winchester, Mass 01890, USA

George Allen & Unwin Australia Pty Ltd,
8 Napier Street, North Sydney, NSW 2060, Australia

First published in 1983

British Library Cataloguing in Publication Data

Tsuboi, Chuji
Gravity.
1. Gravity
I. Title
537′.14 QB334
ISBN 0-04-551072-5
ISBN 0-04-551073-3 Pbk

Library of Congress Cataloging in Publication Data

Tsuboi, Chūji, 1902–
Gravity.
Translation of: Jūryoku. 1979.
1. Gravity. I. Title.
QB331.T7313 1983 531′.14 82-11543
ISBN 0-04-551-073-3 (pbk.)

Set in 10 on 12 point Times by Preface Ltd, Salisbury, Wilts.
and printed in Great Britain by
Butler and Tanner Ltd, Frome and London.

Preface

This book has been written to provide its readers with fundamental knowledge about gravity, more particularly about the use of gravity distribution for understanding the geophysical structure of the Earth. For this purpose, amongst others, the author has attempted to explain as clearly as possible how to apply the elementary potential theory to various practical gravity problems, with the hope that those readers who are not quite familiar with this kind of approach will be able to use it without difficulty. Many more pages have been devoted to this subject in this book, compared with other books dealing with gravity.

The subject of gravity may be regarded as being situated at the centre of a triangle formed by three points representing experimental physics, mathematics of the potential, and the internal structure of the Earth. Sound combinations of any two or all of these subjects are possible only when they are connected by way of the point representing gravity.

The book is fundamentally based on the lectures on gravity which the author gave at the University of Tokyo for many years, with several additional results which he has obtained after his retirement from the university.

CHUJI TSUBOI
October 1981

Contents

List of tables

1 Gravity

1.1 What is gravity?

Every material body around us will begin to move towards the ground when it is released freely from a state of rest. This motion is caused by a force acting on the body and this force is called gravity. The initial direction of the falling motion is said to be vertically downward at that point, and the direction that is at a right angle to the vertical is said to be horizontal.

When you hold a body in your hand, you will feel that your hand is pushed down by its weight. This sense of weight is caused by the gravity force acting on the body. The length of a rubber band will be increased when its upper end is fixed and a body is hung at its lower end. This is also the effect of the gravity force acting on the body.

The force of gravity is acting not only on a body at rest but also on a body in motion. The downward speed of a body which is released from a state of rest increases with time, but the upward speed of a body, when it is thrown high, decreases with time, because the gravity force is continuously acting downward on the bodies irrespective of their directions of motion. If a body is thrown horizontally, its horizontal speed is not affected by gravity at all, but its height decreases with time in the same way as when it was released from a state of rest.

The force of gravity acting on a body increases in direct proportion to its mass m. If this proportional constant is written g, the force is mg in magnitude. Taking the z-axis downward, the equation of motion of a freely falling body is

$$m \frac{d^2z}{dt^2} = mg \tag{1.1}$$

with the conditions

$$\frac{dz}{dt} = 0 \quad \text{and} \quad z = 0$$

at $t = 0$. Hence it is seen that

$$\frac{d^2z}{dt^2} = g \tag{1.2}$$

$$\frac{dz}{dt} = gt \tag{1.3}$$

$$z = \tfrac{1}{2}gt^2. \tag{1.4}$$

This shows that the downward acceleration of the body does not depend on its mass nor on its speed. The value of g represents the magnitude of this acceleration.

The unit of acceleration in the cgs system is 1 cm s^{-2} and this unit is called 1 gal, in memory of Galileo Galilei (1564–1642) who was a pioneer in the study of the motion of bodies falling under gravity. The value of g is about 980 cm s^{-2} (or 980 gal) near the Earth's surface. One thousandth of a gal is called 1 milligal (mgal) which is equal to 10^{-3} cm s^{-2}; for example the difference between 980.123 gal and 980.120 gal is said to be 3 mgal. This unit mgal is very often and conveniently used in geophysical problems. Until the 1940s, this was the limit of accuracy attainable in the measurements of the value of g. Recent progress in measuring techniques of g has made it possible for us to achieve far greater accuracy. One thousandth of 1 mgal has come to be used in some special problems related to gravity and this is called 1 microgal (μgal, 10^{-6} gal) which is equal to 10^{-6} cm s^{-2}.

The magnitude of the gravity force acting on a mass of m g is mg dyne and that acting on a unit mass of 1 g is g dyne. Therefore the magnitude of the gravity force acting on a mass of 1 g and that of downward acceleration of a falling body (g gal) are numerically the same, if the same unit system is used. The downward velocity which a body will have one second after it is released from a state of rest is g cm s^{-1}. Twice the distance of free fall of a body within one second after it is released is g cm. All these four quantities have the same numerical values, although they differ dimensionally.

acceleration due to gravity:	g cm s^{-2}	(L/T^2)
force of gravity acting on 1 g:	g dyne	(ML/T^2)
downward velocity of free fall in 1 s:	g cm s^{-1}	(L/T)
twice the distance of free fall in 1 s:	g cm	(L)

In order to be exact in using these units, we should say, for instance:

the acceleration due to gravity at this point is 980.123 gal

or

the magnitude of gravity force acting on one gram at this point is 980.123 dyne.

But in many practical cases, we often simplify the statements and use expressions such as:

gravity at this point is 980.123 gal

or even omitting the name of the unit,

gravity at this point is 980.123.

These are by no means exact statements, but will not produce any practical harm if they are used with prior understanding. In fact, these simple expressions will frequently be used in this book. Throughout this book, the cgs system of units is used and in many cases the names of units and their dimensions will be omitted as long as no misunderstanding is liable to occur. For example, the density of sea water which is 1.03 g cm^{-3} will be simply said to be 1.03.

1.2 Attraction of the Earth

The greater part of the force of gravity acting on a body above the Earth's surface is caused by the attraction of the Earth. Other than the attraction, a centrifugal force is also acting on the body due to the revolution of the Earth around its axis. This centrifugal force is greatest at the Equator but even there, it is only about 1/300 of gravity. Forgetting this small centrifugal force for the time being, let us see here what relationship exists between the value of g and the mass of the Earth, M, using the following simplifying assumptions:

(a) The Earth is perfectly spherical in shape;
(b) The density distribution within the Earth is spherically symmetric;
(c) the Earth is at rest.

Now let us express the radius of the Earth (6.4×10^8 cm) by R, and the universal constant of attraction by G (some people use k^2 instead of G). The value of G represents the magnitude of the force of attraction acting between two material points, each 1 g in mass, which are placed at a distance of 1 cm. From the inverse square law of attraction,

$$g = GM/R^2 \tag{1.5}$$

or

$$M = \frac{980 \times (6.4 \times 10^8)^2}{G}.$$

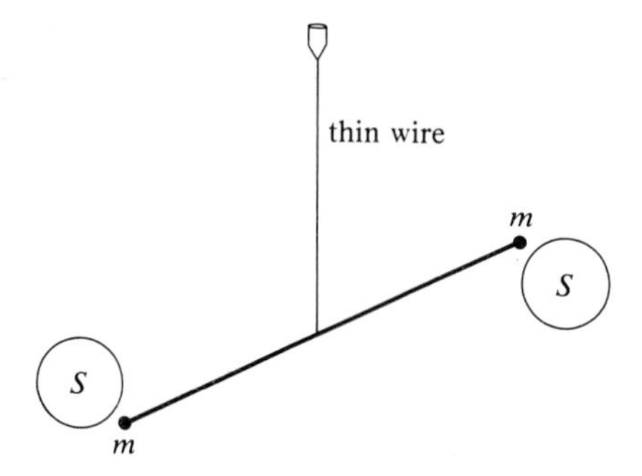

Figure 1.1 An arrangement for measuring the universal constant of attraction. Attractions between m and S cause a torsion of the wire.

If the value of the constant G is known, then the mass of the Earth M can be calculated by means of this relation. The value of G can be found experimentally by an arrangement which is shown diagrammatically in Figure 1.1. A straight rod, to which two masses m are attached at the ends, is suspended horizontally by a thin elastic wire from above and is kept horizontal. Now bring two large masses S close to the two masses m as shown in the figure. The attraction forces acting between S and m will cause the rod to rotate in the horizontal plane and the suspension wire will be twisted a little, until the elastic torque produced in the wire by this rotation will just balance with the moments of force around the central axis due to the attractions between S and m. By measuring this small angle of rotation of the rod and combining it with instrumental constants, the value of G can be obtained.

Measurements of this sort have been made by several investigators, the first of which, by H. Cavendish (1731–1810), is historically famous. The up-to-date value of G is

$$G = 6.672 \times 10^{-8}\ \text{cm}^3\ \text{g}^{-1}\ \text{s}^{-2}.$$

This numerical value of 6.67 is accidentally very close to 20/3 and this is convenient for us when we make approximate numerical calculations related to gravity problems.

Now that the value of G is known, the mass of the Earth M can be calculated as follows:

$$\begin{aligned} M &= \frac{gR^2}{G} \\ &= \frac{980 \times (6.4 \times 10^8)^2}{6.67 \times 10^{-8}} \\ &\approx 6.0 \times 10^{27}\ \text{g}. \end{aligned} \tag{1.6}$$

Dividing this mass by the volume of the Earth, $\frac{4}{3}\pi R^3$, the mean density $\bar{\rho}$ of the Earth is found to be

$$\begin{aligned}\bar{\rho} &= \frac{gR^2}{G} \bigg/ \frac{4}{3}\pi R^3 \\ &= \frac{3g}{4\pi GR} \\ &\approx 5.5. \qquad (1.7)\end{aligned}$$

It will be interesting here to get rough ideas about the strength of attraction forces between two masses around us. The attraction working between two persons, each 60 kg in mass, sitting at a distance of five metres is approximately 0.001 dyne, which is as small as 1/60 000 000 000 of the weight of each person. A lead sphere with a radius of 3.2 m exerts an attraction of 0.001 dyne to a mass of 1 g placed on its surface. Again, an infinite horizontal plate of rock with a density of 2.7 and a thickness of 8.9 m exerts an attraction of 0.001 dyne to a mass of 1 g, irrespective of its distance from the plate.

The gravity force at a point depends on its height above sea level. If we go higher, the distance from the centre of the Earth increases and the force of attraction of the Earth decreases, according to the inverse square law of attraction. The force of attraction F of the Earth at a height h above sea level is

$$\begin{aligned}F &= \frac{GM}{(R+h)^2} \\ &\approx \frac{GM}{R^2}\left(1 - \frac{2h}{R}\right) \\ &\approx g - \frac{2g}{R}h, \qquad (1.8)\end{aligned}$$

provided that h is very small compared with R. In this approximation, F decreases linearly with increasing h at a rate

$$\frac{2g}{R} = 0.3086 \text{ mgal m}^{-1}. \qquad (1.9)$$

So the value of g decreases at a rate of about 0.3 mgal at every metre elevation, or 1 mgal at every 3.2 m elevation. Thus at the summit of Mt Fuji in Japan ($h = 3776$ m), the value of g is 1165 mgal (1.165 gal) less than the value of g at zero height ($h = 0$).

1.3 Centrifugal force due to the revolution of the Earth

The attraction of the Earth is not the whole part of gravity. The Earth is revolving around its axis and everything fixed to its surface is acted on by the centrifugal force. This force must be taken into consideration in studying gravity problems as we see them from the co-ordinate system which is fixed to the revolving Earth. This co-ordinate system has an acceleration with respect to absolute space and in that sense it is a non-inertial system. By introducing the centrifugal force, it becomes possible for us to regard our revolving system as an inertial system.

The Earth makes one revolution around its axis in one siderial day (86 164 s), the angular velocity ω being

$$\begin{aligned}\omega &= 2\pi/86\,164 \text{ rad s}^{-1}\\ &\approx 7.3 \times 10^{-5} \text{ rad s}^{-1}.\end{aligned} \tag{1.10}$$

If the Earth were perfectly spherical in shape, the centrifugal force f at a point of latitude ϕ due to this angular velocity would be

$$f = \omega^2 R \cos\phi. \tag{1.11}$$

The component of this centrifugal force in the direction of the Earth's attraction is nearly equal to $\cos\phi$ times f, namely $\omega^2 R \cos^2\phi$ and its effect is to decrease the effect of attraction. At the Equator, $\phi = 0$, the centrifugal force is the largest and is

$$\begin{aligned}\omega^2 R \cos^2\phi &= \omega^2 R\\ &= 3.4.\end{aligned} \tag{1.12}$$

This is about 1/300 of g itself. No centrifugal force is working at the

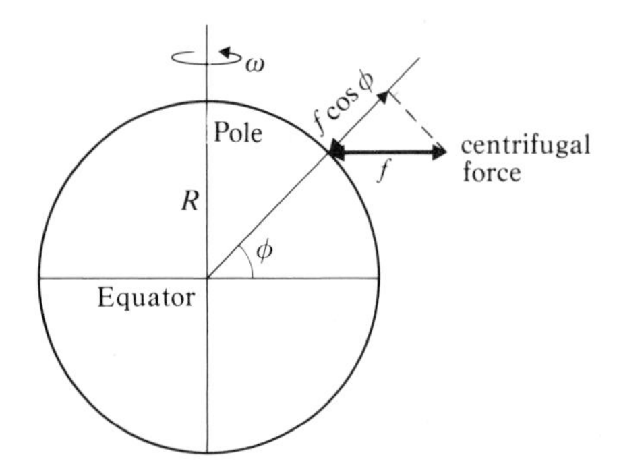

Figure 1.2 The centrifugal force due to the revolution of the Earth.

North and South Poles where $\phi = \pi/2$. According to the results of recent measurements of g;

$$g \text{ at the Poles} = 983.218$$
$$g \text{ at the Equator} = 978.032$$

the difference being 5.2. This value does not agree with the value of 3.4 calculated above. This disagreement implies that the real Earth is not spherical in shape as it was assumed to be in the preceding calculations.

1.4 Attractive force and centrifugal force

Suppose a very high tower is built on the Equator. If the height of the tower is h, the centrifugal force f acting at the top is

$$\begin{aligned} f &= \omega^2(R + h) \\ &= \omega^2 R\left(\frac{R + h}{R}\right) \end{aligned} \tag{1.13}$$

and it increases linearly with h as shown by the broken line in Figure 1.3. On the other hand, the attractive force F of the Earth at the top of the tower is

$$\begin{aligned} F &= \frac{GM}{(R + h)^2} \\ &= \frac{GM}{R^2}\left(\frac{R}{R + h}\right)^2 \end{aligned} \tag{1.14}$$

and this decreases with h as the line curve in Figure 1.3 shows. The

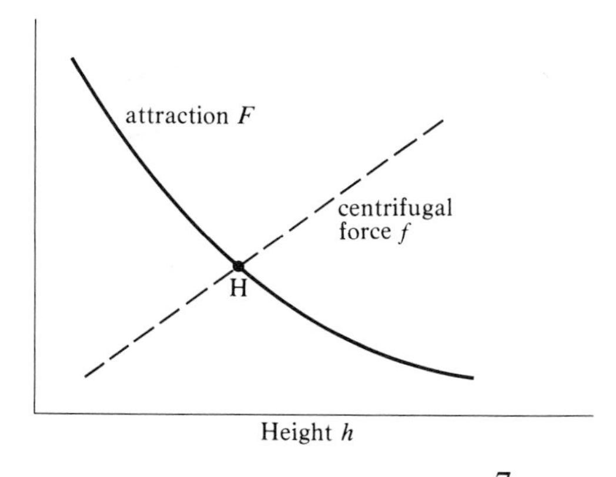

Figure 1.3 Dependence of attraction (line) and centrifugal force (dotted line) on height.

curves for f and F in Figure 1.3 intersect at a point H. If the tower is lower than H, the attraction F acting at the top towards the Earth is stronger than the centrifugal force f acting away from the Earth, so that a body will move towards the Earth when it is released at the top of the tower. On the contrary, if the tower is higher than H, the centrifugal force is stronger than the attraction of the Earth, so that the body released at the top will move away from the Earth. At the very height that corresponds to the point H, the two forces balance each other, so that the body released will move neither towards nor away from the Earth, but will keep revolving around the Earth at the rate of one revolution a day and will remain above the same point of the Equator. This is simply a stationary artificial satellite. The height h of the point H can be calculated by equating the attractive force F and the centrifugal force f at that point:

$$\omega^2 R\left(\frac{R+h}{R}\right) = \frac{GM}{R^2}\left(\frac{R}{R+h}\right)^2 \tag{1.15}$$

or

$$\left(\frac{R+h}{R}\right)^3 = \frac{g}{\omega^2 R}.$$

By inserting the numerical values of R, g and ω, the value of h is found to be

$$h = 5.6\,R = 36\,000 \text{ km}.$$

All the stationary artificial satellites have this fixed height.

1.5 Gravity and latitude

The force of gravity at a point on the Earth is the vectorial resultant of the attraction of the Earth and the centrifugal force. Let us now do some calculations to see how the value of g varies with latitude. For simplicity, the Earth is assumed to be perfectly spherical in shape and revolving around its axis with an angular velocity ω.

Take a triangle that is formed by three sides, g, F and f as in Figure 1.4. For this triangle,

$$\begin{aligned} g^2 &= F^2 + f^2 - 2Ff\cos\phi \\ &= \left(\frac{GM}{R^2}\right)^2 + (\omega^2 R\cos\phi)^2 - \frac{2GM}{R^2}\,\omega^2 R\cos^2\phi \end{aligned} \tag{1.16}$$

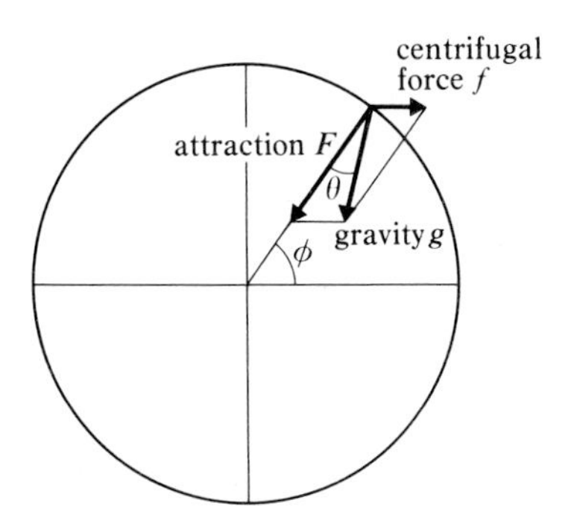

Figure 1.4 Attraction, centrifugal force and gravity on a spherical Earth.

hence

$$g \approx \frac{GM}{R^2}\left(1 - \frac{\omega^2 R}{GM/R^2}\cos^2\phi\right). \tag{1.17}$$

As a first rough approximation, if we put

$$\frac{GM}{R^2} = 980$$

and

$$\frac{\omega^2 R}{GM/R^2} = \frac{1}{300},$$

we obtain

$$g = 980(1 - 0.0033\cos^2\phi)$$

$$= 977(1 + 0.0032\sin^2\phi) \tag{1.18}$$

This g value changes with latitude as shown by the curve in Figure 1.5.

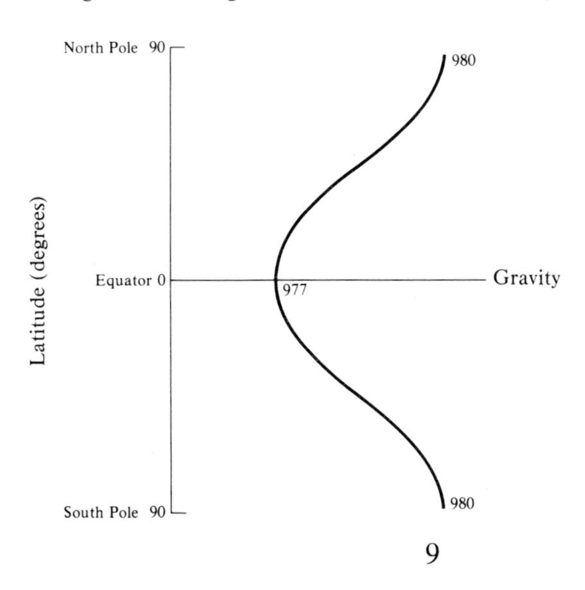

Figure 1.5 Variation of gravity with latitude on a simplified Earth.

This curve does not fit observed values well, because of the very rough approximations used in the above calculations. But this simple calculation shows the general picture of the variation of g according to latitudes.

1.6 The geoid

The above calculations based on the assumption that the Earth is perfectly spherical yield some peculiar results. One of these is that the direction of g so calculated is not at a right angle to the surface of the Earth, or, in other words, that the surface of the Earth is not horizontal everywhere. As seen from Figure 1.4, there is a geometrical relation

$$\frac{g}{\sin\phi} = \frac{f}{\sin\theta}, \tag{1.19}$$

and the angle θ of the slope can be calculated from this. Putting

$$f = \omega^2 R \cos\phi$$

in the above relation, we obtain

$$\begin{aligned} \theta &= \frac{f \sin\phi}{g} \\ &= \frac{\omega^2 R \cos\phi \sin\phi}{g} \\ &= \frac{1}{2}\frac{\omega^2 R}{g}\sin 2\phi \end{aligned} \tag{1.20}$$

if θ is small. From this, the angle of slope θ is seen to be largest at $\phi = 45°$ and is $5'$ in angle. This angle corresponds to an inclination of 1.5 m in a distance of 1 km. In order to avoid such an unexpected conclusion, we have to consider what relationship should exist between the direction of gravity and the Earth's surface, and we have to define the shape of the Earth based on these considerations. The definition of the shape of the Earth which will be adopted is as follows:

> the shape of the Earth's surface is such that it is perpendicular to the direction of gravity at every point;

or

> the shape of the Earth's surface is such that, at every point, it is perpendicular to the vectorial resultant of the attraction of the Earth and the centrifugal force;

or

> the surface of the Earth is horizontal everywhere.

The relationship between the directions of the surface and the three forces (attraction, centrifugal force and gravity) is seen in Figure 1.6. This surface, which is horizontal everywhere, is called the geoid. The geoid is a theoretical smooth surface and is different from the irregular physical surface of the actual Earth; very simply, the shape of the geoid is that which the surface of sea water would take were it to cover the whole surface of the Earth. This will be made theoretically possible by considering that a network of narrow canals was built over the continents and the sea water was led into them.

Since the geoid is horizontal everywhere, no work is needed to be done by the gravity force when a body is moved along the geoid. In dynamics, we say that such a surface has a constant gravity potential or is an equipotential surface. In general, there can be as many equipotential surfaces as the value of the potential assigned to each of them, and no two of them intersect. When we simply say 'the geoid' in gravity problems, it means the equipotential surface that coincides with the ocean surface on average.

Imagine that the solid part of the Earth is a perfectly rigid sphere, covered by sea water and the whole system is making a revolution around its axis. In that case, the sea water near the Equator will bulge out because of the centrifugal force acting on its mass. The sea would then be much deeper near the Equator than at higher latitudes. Of course, this does not actually take place. This is because the 'solid' part

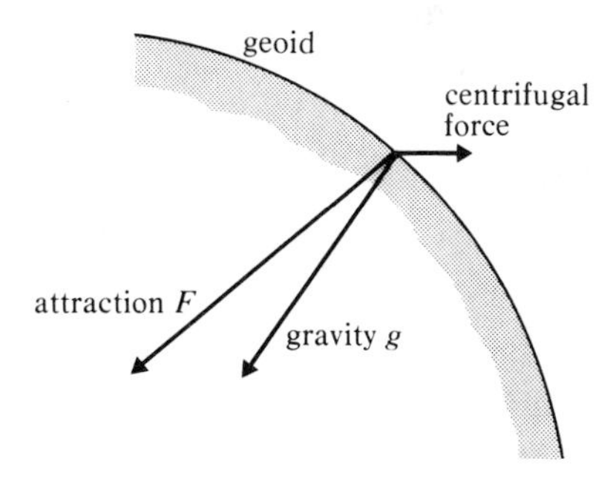

Figure 1.6 Relationship between the geoid, attraction, centrifugal force and gravity. Gravity is a vectorial resultant of attraction and centrifugal force. The geoid is perpendicular to gravity.

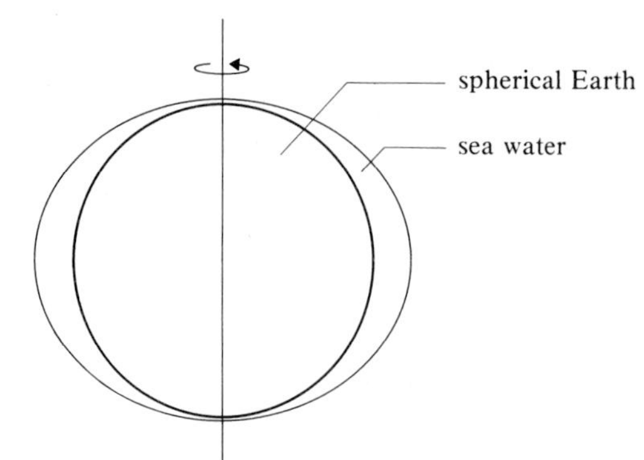

Figure 1.7 If the Earth were perfectly spherical, and covered entirely by water, the oceans would be deepest along the Equator.

of the Earth is not really solid and in the billions of years since its formation, the 'solid' part has behaved as a liquid and has been deformed to attain a hydrostatic equilibrium itself.

1.7 The Eötvös effect

Let us see what will happen to the gravity force acting on a body when it is moving along the Earth's surface. If the body is moving eastward relative to the Earth, its angular velocity around the Earth's revolution axis will be increased, and consequently the centrifugal force acting on it will also be increased. Conversely, if the body is moving westward, the angular velocity will be decreased, and consequently the centrifugal force will also be decreased. If the body is moving at the Equator with an eastward velocity v relative to the Earth, its velocity will increase from its original value ωR ($7.3 \times 10^{-5} \times 6.4 \times 10^3 = 0.5$ km s^{-1}) to $(\omega R + v)$. Consequently, the centrifugal force will be increased from $(\omega R)^2/R$ to $(\omega R + v)^2/R$. If $v \ll \omega R$, the difference between the two is

$$\frac{(\omega R + v)^2}{R^2} - \frac{(\omega R)^2}{R^2} \approx 2v\omega. \tag{1.21}$$

If the walking speed of a man is 1 m s^{-1}, for instance,

$$2v\omega = 2 \times 10^2 \times 7.3 \times 10^{-5}$$

$$\approx 15 \times 10^{-3}.$$

The apparent decrease in the value of g acting on the man will be 15 mgal. If the weight of the walking man is 60 kg, he would feel as if his weight were decreased by

$$60 \times 10^3 \times 15 \times 10^{-3} \div 980 = 0.9 \text{ g}.$$

A situation like this is not without relevance to gravity problems and comes to be an important practical problem when gravity values over the sea are to be measured on board a moving ship. For precise measurements of gravity, the east–west velocity of the ship should be known, and corrections should be made for it in calculating the value of g. The correction for this effect is 15 mgal for a velocity of 1 m s^{-1} at the Equator. The velocity of the ship should be known with an accuracy of $100/15 = 7$ cm s^{-1}, if the accuracy of g is 1 mgal. This accuracy corresponds to 250 m h^{-1} or 0.14 knots. Such a precise measurement of the velocity of a ship has been almost impossible until recently, when the technique of artificial satellite tracking was successfully introduced for this purpose.

The influence of the east–west velocity of a body upon the apparent gravity value was first pointed out by a Hungarian physicist, R. v. Eötvös (1847–1919), and is now called the Eötvös effect.

Making use of this Eötvös effect, it will be possible, at least theoretically, to determine the direction of north–south without referring to the stars or to the direction of a magnetic needle. Build a long-period balancing pendulum as shown in Figure 1.8 and place it on a table which is made to rotate around its vertical axis. By this rotation, the two masses attached to the ends of the pendulum rod move in opposite directions. If the rod is in the north–south direction at an instant and if the table is turning clockwise as seen from above, the southern mass moves westward and the northern one eastward. The gravity force acting on the southern mass is stronger than that working on the

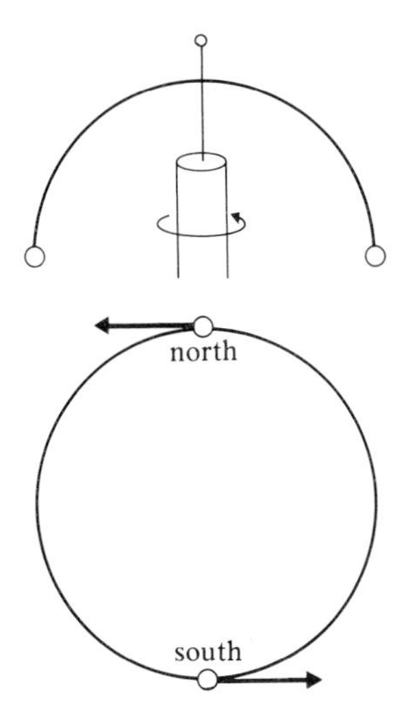

Figure 1.8 A balancing pendulum placed on a rotating table. The pendulum will tilt down to the north if the table is rotating counterclockwise.

northern mass according to the Eötvös effect, and a moment of force will act on the rod which will make it tilt down towards the south. If the rotation period of the turning table is made equal to the free-oscillation period of the rod pendulum, the inclination of the rod will resonate with the change of the moments of force acting on it and the inclination will grow large enough to be observed.

In the above calculations about the Eötvös effect, it was assumed that $v \ll \omega R$. If v becomes very large, this approximation is no longer acceptable. In the extreme case that v is so large that the centrifugal force becomes equal to the force of attraction of the Earth, the two forces will balance each other and the apparent gravity force working on a body will be zero. If a body is shot out horizontally with this velocity along the Equator, it will not fall to the ground but will fly around the Earth skimming its surface. The value of such v is given by the relation

$$g = \frac{(\omega R + v)^2}{R}. \tag{1.22}$$

From this, we obtain

$$\begin{aligned} \omega R + v &= \pm\sqrt{(gR)} \\ &= \pm 7.9 \text{ km s}^{-1}. \end{aligned}$$

A point on the Equator already has an eastward velocity of 0.5 km s^{-1} by the revolution of the Earth, so that a body which is shot eastward with a velocity:

$$7.9 - 0.5 = 7.4 \text{ km s}^{-1}$$

or one that is shot westward with a velocity:

$$7.9 + 0.5 = 8.4 \text{ km s}^{-1}$$

relative to the Earth will become an artificial satellite, which flies around the Equator skimming its surface. The time needed for this satellite to complete one revolution around the Earth is

$$\begin{aligned} T &= \frac{2\pi R}{\omega R + v} \\ &= \frac{2\pi R}{\sqrt{(gR)}} \end{aligned}$$

$$= 2\pi\sqrt{\frac{R}{g}}. \qquad (1.23)$$

This is equal to the period of a hypothetical simple pendulum of length R swinging under a uniform gravity of 980. The value of this T is 5080 s or about 1 h 25 min.

In this connection, some comments will be made about what is called the weightlessness condition in an artificial satellite. An artificial satellite is such a body that the force of gravity and the centrifugal force acting on it always balance each other. The spaceship itself, as well as an astronaut riding in it, are placed under such a condition and there is no dynamical interaction working between them, so that the astronaut has no reason to feel his weight. This is the state called weightlessness.

The state of weightlessness will occur not only in an artificial satellite. Suppose you are put in a large box which is suspended at a height by a rope. If the rope is broken suddenly, the box and yourself will begin to fall freely together towards the ground in the same way. There is no dynamical interaction between your body and the box so that you have no reason to feel your own weight. This is also the weightlessness condition. Even if the effect is not that large, you will experience a condition more or less similar to weightlessness when you ride in a lift. When the lift begins to move down, you will feel as if your weight has decreased and when it begins to move up, you will feel as if your weight has increased. It is interesting to put a mass on a spring balance, take the apparatus into a lift and watch how the scale reading of the balance changes when it begins to move up or down. With a vertical acceleration, the apparent gravitational force acting on a mass changes.

1.8 Gravity within the Earth

First, the gravity distribution within a spherical Earth with a uniform density ρ will be calculated. The attraction at a point P within the Earth, at a distance r from the centre, is produced by the mass lying deeper than that point, the mass of the spherical shell that lies shallower exerting no attraction. The mass lying deeper than P is $\frac{4}{3}\pi r^3\rho$. The attraction at P due to this mass is therefore

$$\frac{4}{3}\frac{\pi G r^3\rho}{r^2} = \frac{4}{3}\pi G\rho r. \qquad (1.24)$$

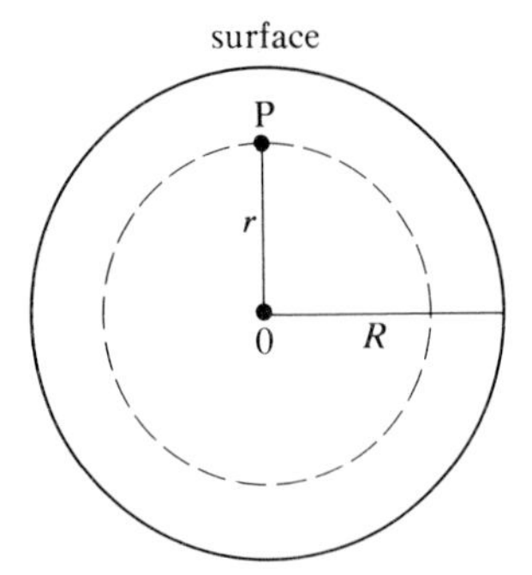

Figure 1.9 Attraction at P is caused by the spherical mass with a radius r.

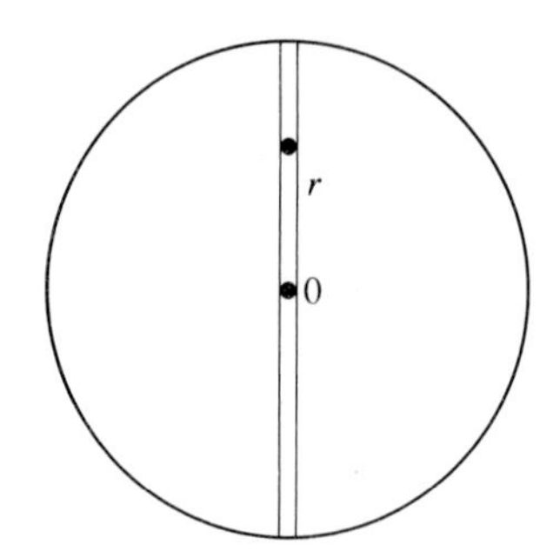

Figure 1.10 A mass dropped into the diametrical well oscillates harmonically with the centre of the Earth as origin.

This attraction is largest on the Earth's surface and decreases linearly with r until it becomes zero at the centre of the Earth. This linear variation of attraction according to its distance from the origin is the same as that which is seen in the case of a simple harmonic oscillator. Imagine that a hole is dug from a point on the surface to its antipode through the centre of the Earth as shown in Figure 1.10. If a mass is dropped into the hole, it will move to the centre of the Earth and after passing the centre will reach the antipodal point where it comes to rest. After that, the mass will move in the opposite direction and repeat the simple harmonic motion. The equation of motion of this mass is

$$\frac{d^2r}{dt^2} = -\tfrac{4}{3}\pi G \rho r$$

and its period of oscillation is

$$T = \frac{2\pi}{\sqrt{(\tfrac{4}{3}\pi G\rho)}} .$$

But since

$$g = \tfrac{4}{3}\pi G \rho R$$

the expression for T will become

$$T = 2\pi\sqrt{\frac{R}{g}} . \tag{1.25}$$

This is the period of a hypothetical simple pendulum with a length R under a uniform gravity of 980 and is also equal to the time needed for

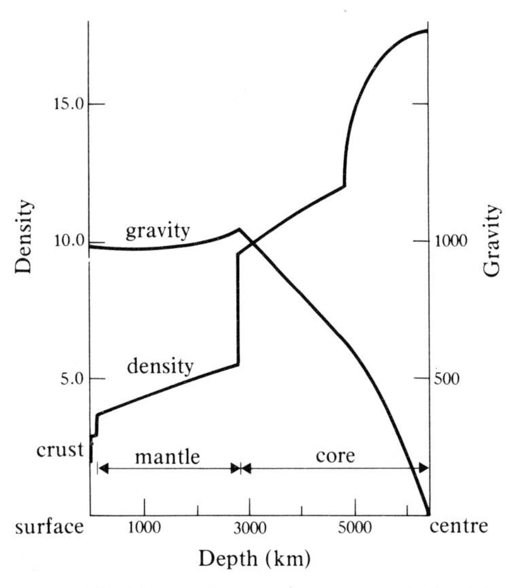

Figure 1.11 Density and gravity within the Earth.

an artificial satellite to go around the Equator, skimming the surface of the Earth.

The density distribution within the actual Earth is far from being uniform, as it has been assumed to be in the above calculations. The distribution of density within the Earth, which has been found by studies of the propagation velocity of earthquake waves through it, is as shown in Figure 1.11. This curve was obtained by Bullen (1936) of Australia. The density changes abruptly at three depths and the four parts of the Earth lying between these depths are called the Earth's crust, mantle, outer and inner cores respectively. In Figure 1.11 the gravity values within the Earth, which are calculated on the basis of this density distribution, are also shown by a curve. If the Earth were uniform in density, the gravity value would decrease linearly with increasing depth. But within the actual Earth, the gravity value remains almost constant down to a depth of about 2500 km, because the effect of the decrease in distance from the centre and that of the increase in density cancel each other out. At a depth of about 3000 km, the value of g is the largest and exceeds 1000 gal.

1.9 Purposes of gravity measurements

The purposes of measuring gravity values over the Earth's surface are manifold. Among these, the following are five main reasons.

1.9.1 To determine the shape of the Earth (Ch. 4)

This has been the classical problem since the days of Isaac Newton. By measuring the values of g at various latitudes, the shape of the Earth can be determined. As will be shown later, the value of g at a latitude ϕ is theoretically given by

$$g_\phi = g_e(1 + \beta \sin^2 \phi - \beta' \sin^2 2\phi), \tag{1.26}$$

where

$$\beta = \frac{5}{2}\frac{\omega^2 a}{g_e} - \varepsilon - \frac{17}{14}\frac{\omega^2 a}{g_e}\varepsilon,$$

$$\beta' = \frac{\varepsilon}{8}\left(\frac{5\omega^2 a}{g_e} - \varepsilon\right),$$

and g_e is the value of g at the Equator, ω the angular velocity of rotation of the Earth, a the equatorial radius of the Earth, b the polar radius and ε the ellipticity of the Earth $(a - b)/a$. Making use of all the observed values of g at a number of points over the Earth, numerical values of the constants in the above formula have been found to be

$$g_\phi = 978.03185(1 + 0.0053024 \sin^2 \phi - 0.0000059 \sin^2 2\phi) \tag{1.27}$$

$$\varepsilon = 1/298.25.$$

Here, the shape of the Earth has been assumed to be an ellipsoid of revolution, so that there is no term included in the formula that depends on longitude.

1.9.2 To find the underground-mass distribution (Chs 7 & 8)

The gravity formula given in the preceding section represents the average distribution of gravity on the whole surface of the Earth. The value of g at a particular point does not necessarily agree with the value calculated by this formula for that point. The difference between the two is called the gravity anomaly. Gravity anomaly values are essential data from which the anomalous underground mass can be calculated. This is the problem which is most actively dealt with in relation to gravity. There are large-scale problems related to the constitution of continents and oceans, or of large mountain ranges and oceanic trenches. The concept of isostasy is one of the outcomes derived from studies of this kind. Smaller-scale problems than these are local geological structures; still smaller problems are those related to exploration of underground natural resources, such as oil.

1.9.3 To estimate the elasticity of the Earth (Ch. 12)

This is concerned with the phenomena that are called the Earth tides. The Earth is acted on by the attractive forces of the Moon and the Sun and, according to their motions, the Earth is periodically deformed. The value of g at a point on the Earth is affected not only by the direct attraction of these celestial bodies but also by the attraction produced by the deformed part of the Earth. Also, the distance of a point on the Earth from its centre changes according to this deformation and therefore the gravity value. From these observations, the elasticity of the Earth can be estimated.

1.9.4 To see if g values change slowly with time (Ch. 12)

This is a relatively new problem in geophysics aimed at solving questions of whether or not some changes take place in mass distribution in the depths of the Earth in relation to occurrences of earthquakes or volcanic eruptions. If some mass movements take place, the gravity values in the area concerned will change and, from these observations it will be possible to calculate the mass movement.

1.9.5 To standardize physical and chemical constants (Ch. 3)

This is not a purely geophysical problem, but is rather related to physics and chemistry. A number of units used in physics and chemistry are defined without referring to the gravity value. For instance, the temperature of 100° C is defined as the temperature at which pure water will boil under a pressure of $P = 1.113250 \times 10^3$ dyne cm^{-2}. But the atmospheric pressure is actually measured by the height h of a mercury-column barometer through the relation

$$P = \rho g h.$$

The value of P cannot be known unless the absolute value of g is known at the point where this measurement is made.

References and further reading

Bullen, K. E. 1936. The variation of density and the ellipticities of strata of equal density within the Earth. *Month. Not. R. Astr. Soc., Geophys. Suppl.* **3**, 395.

Cavendish, H. 1798. Experiments to determine the density of the Earth. *Phil. Trans R. Soc., Lond.* **88**, 469.

Heyl, P. R. 1930. A redetermination of the constant of gravitation. *J. Res. US Bur. Stand.* **5**, 1243.

Heyl, P. R. and P. Chrzanowski 1942. A new determination of the constant of gravitation. *J. Res. US Bur. Stand.* **29**, 1.

Lambert, W. 1939. Density, gravity, pressure and elasticity in the interior of the Earth. In *Physics of the Earth*, vol. VII, 329. New York: McGraw-Hill.

Lambert, W. and F. W. Darling 1951. *Internal constitution of the Earth*. New York: Dover.

Poynting, J. H. 1894. *The mean density of the Earth.* London: Griffin.

2 Measurements of gravity

2.1 Absolute and relative measurements

There are two different types of measurement of gravity values – the absolute measurement and the relative measurement. In the absolute measurement, the value of g at a place is found by operations made at that point only, without referring to gravity values at other places. In the relative measurement, on the other hand, it is the difference in the values of g between two places that is measured.

One of the methods for finding the absolute value of g is to swing a simple pendulum and measure its period of oscillation. A simple pendulum is a pendulum in which a dimensionless material point is suspended by a perfectly flexible and unstretchable massless string of length l. When this is swung in a vertical plane with a very small amplitude, its period of oscillation T is given by

$$T = 2\pi\sqrt{\frac{l}{g}}. \tag{2.1}$$

By measuring T and l, the value of g can be calculated according to this expression.

This is very simple in theory, but difficulties lie in actualising any ideal pendulum and in finding its real length l. All that can be actually done is to suspend a heavy mass of a finite size by a wire which has mass and elasticity. In order to reduce the results obtained with this real pendulum to the case of the ideal simple pendulum, many elaborate supplementary measurements are needed, so that the whole calculation is a complex and lengthy procedure. If an accuracy of 1 mgal is aimed at in the measurement of g, the length l of the pendulum while it is swinging must be known to an accuracy of 1 micron (10^{-3} mm), in the case when its length l is 1 m. This is very difficult to achieve.

There are several other methods which have been developed for the purpose of accurate absolute measurements, for example, by using a reversible pendulum or a falling body. But unfortunately, they need bulky apparatus and elaborate manipulations, as will be explained later. It is therefore not feasible to try to make absolute measurements at each of a number of places where gravity values are to be found.

The relative measurement, on the other hand, is much simpler. One of the methods for this purpose is to swing one and the same rigid pendulum, first at a place A and then at another place B and to compare the periods at these places. The periods of the pendulum at the two places, A and B, are

$$T_A = 2\pi\sqrt{\frac{K}{g_A}} \qquad T_B = 2\pi\sqrt{\frac{K}{g_B}}. \tag{2.2}$$

K is the physical constant of the pendulum and is the same at A and B. From these equations, it follows that

$$T_A^2 : T_B^2 = \frac{1}{g_A} : \frac{1}{g_B} \tag{2.3}$$

or

$$g_A - g_B \approx 2 \times 980 \times (T_B - T_A)$$

if $T_A \approx T_B \approx 1$ s. By this relation, the difference between g_A and g_B can be found without knowing the value of the constant K.

Other than rigid physical pendulums, static gravimeters of several types have been developed for the purpose of making relative measurements of gravity, and are now widely used. In a gravimeter, a mass is fixed to the end of an elastic spring and the force of gravity acting on it and the elastic force of the spring produced by its elongation are balanced. By measuring changes in elongation of the spring at various places, gravity differences between these places can be found.

Whether the devices are rigid pendulums or elastic springs, they are similar in that it is the difference in the value of g between two places that can be measured. In order to find the values of g at a number of points, relative measurements alone are not sufficient. An absolute measurement must be made at at least one of the points, and all other points should be tied either directly or indirectly to this standard point by relative measurements. If g_A is the value of g at a point A determined by an absolute measurement, the value of g_B at another point B can be found from the relation

$$g_B = \underbrace{(g_B - g_A)}_{\text{relative measurement}} + \underbrace{g_A}_{\text{absolute measurement}} \tag{2.4}$$

2.2 Borda's pendulum

The pendulum measurement which was made by J. C. Borda (1733–1799) for absolute determination of g about two hundred years ago is

Figure 2.1 Borda's pendulum (1790) used for gravity measurements.

historically famous. He built a bulky apparatus as shown in Figure 2.1, suspended a heavy metallic ball by a wire and elaborately measured its length and period of oscillation. Such a pendulum as this is far from being an ideal simple pendulum and results of present-day precision cannot be obtained by this arrangement, although it is still useful for educational demonstrations.

2.3 Reversible pendulum

The difficulty of actualizing the length of a simple pendulum was overcome by Kater (1818) who introduced the idea of the reversible pendulum. A reversible pendulum is a rigid-rod pendulum of a mass M as shown in Figure 2.2, which can be swung around two axes fixed at distances h_1 and h_2 from its centre of gravity. Generally, when a rod

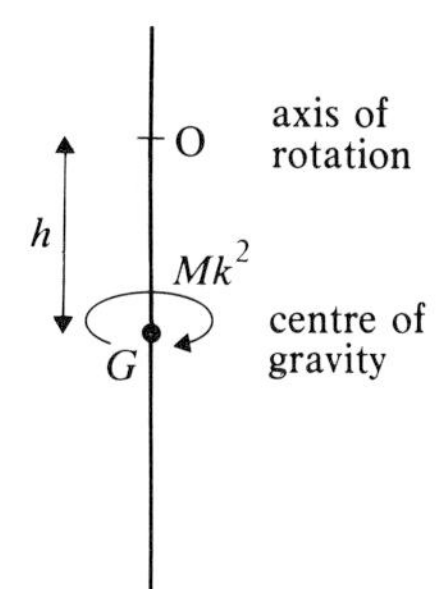

Figure 2.2 A rod pendulum where O represents the axis of oscillation, G the centre of gravity, h the distance between O and G, and Mk^2 the moment of inertia of the rod around O.

pendulum is swung around an axis which is at a distance h from its centre of gravity, its period of oscillation T is given by

$$T = 2\pi\sqrt{\frac{M(k^2 + h^2)}{Mgh}} \tag{2.5}$$

where Mk^2 is the moment of inertia of the rod around a horizontal axis passing through its centre of gravity. Imagine now a simple pendulum, the period of oscillation of which is exactly equal to the period of this rigid pendulum. If the length of this hypothetical simple pendulum is l, then

$$T = 2\pi\sqrt{\frac{l}{g}} = 2\pi\sqrt{\frac{M(k^2 + h^2)}{Mgh}}, \tag{2.6}$$

and from this, we obtain

$$h^2 - lh + k^2 = 0. \tag{2.7}$$

Regarding this as an equation for the unknown h, the solution is

$$h = \tfrac{1}{2}\{l \pm \sqrt{(l^2 - 4k^2)}\}.$$

There are two values h_1 and h_2, both of which satisfy the above equation:

$$h_1 = \tfrac{1}{2}\{l + \sqrt{(l^2 - 4k^2)}\}$$

$$h_2 = \tfrac{1}{2}\{l - \sqrt{(l^2 - 4k^2)}\}$$

and there is a relation

$$h_1 + h_2 = l.$$

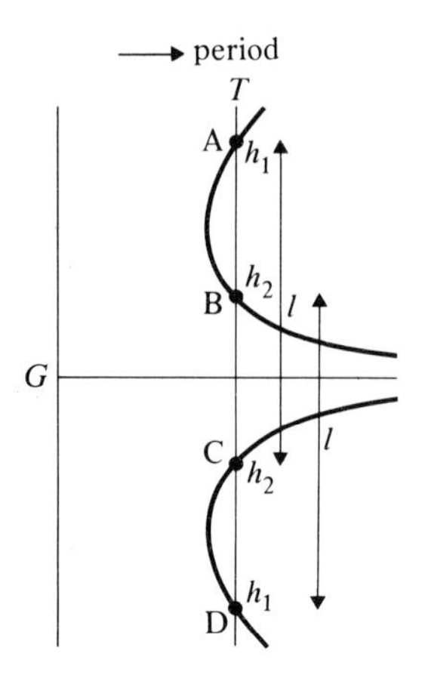

Figure 2.3 Variation of the period of oscillation according to the distance of the oscillation axis from the centre of gravity G. A, B, C and D give the same periods. The distance between A and C, or B and D is equal to the length l of a simple pendulum which oscillates with this period.

The curve in Figure 2.3 shows how the period of the rod pendulum changes when the oscillation axis is shifted along the central line of the rod. As can be seen from this curve, there are four points, A, B, C and D on it, two on each side of the centre of gravity, which give the same period of oscillation. The distance between A and C or between B and D is $(h_1 + h_2)$ and can be measured accurately. From the relation

$$T = 2\pi\sqrt{\frac{l}{g}} = 2\pi\sqrt{\frac{(h_1 + h_2)}{g}}, \tag{2.8}$$

the value of g can be found from

$$g = 4\pi^2 \frac{(h_1 + h_2)}{T^2}. \tag{2.9}$$

This means that, if by careful adjustments of the positions of the two axes, the periods of oscillation around them can be made equal, then the distance between the two axes will give the length l of a simple pendulum which has the same period as this physical pendulum. It is important to notice that the length $l = h_1 + h_2$ can be found without knowing h_1 and h_2 separately. The pendulum will be reversed upside down when it is suspended at A and C, or at B and D. Because of this reversal, it is called a reversible pendulum.

It is, however, an extremely laborious task to shift one or both of the axes until the periods of oscillation around them become exactly equal and then to measure the distance between them. It is much simpler to fix the two axes from the outset and to attach a moveable small mass near one end of the pendulum as shown in Figure 2.4. This mass is screwed in or out until the periods around A and C become equal. Then, by

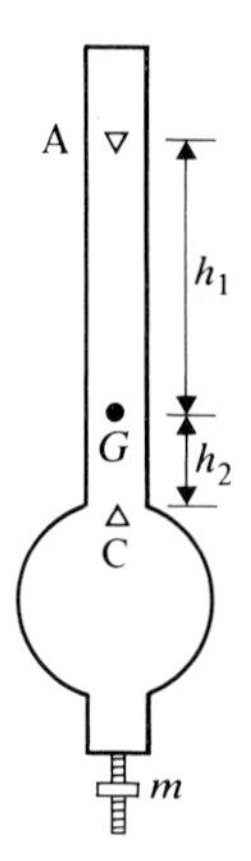

Figure 2.4 A reversible pendulum. The periods of oscillation around A and C are adjusted to be equal by adjusting the position of a small mass m.

measuring the period and combining it with the fixed value of $(h_1 + h_2)$, the value of g can be known.

Even if this is done, it is still difficult to make the two periods around A and C precisely equal. Now let the period of the pendulum around A be T_1 and that around C be T_2, the difference $(T_1 - T_2)$ being very small. Since

$$T_1 = 2\pi\sqrt{\frac{k^2 + h_1^2}{gh_1}}, \qquad T_2 = 2\pi\sqrt{\frac{k^2 + h_2^2}{gh_2}}, \tag{2.10}$$

the period T of the simple pendulum that has the length $(h_1 + h_2)$ is given by

$$\begin{aligned} T^2 &= \frac{4\pi^2}{g}(h_1 + h_2) \\ &= \frac{T_1^2 + T_2^2}{2} + \frac{T_1^2 - T_2^2}{2}\frac{h_1 + h_2}{h_1 - h_2}, \end{aligned} \tag{2.11}$$

or by

$$T = \frac{T_1 + T_2}{2} + \frac{T_1 - T_2}{2}\frac{h_1 + h_2}{h_1 - h_2}, \tag{2.12}$$

neglecting $(T_1 - T_2)^2$ which is a very small quantity. The second terms on the right-hand sides of the expressions are small corrections to be added to the mean value of T_1^2 and T_2^2 in order to find T^2 or to the mean value of T_1 and T_2 in order to find T. Although h_1 and h_2 which are not known are contained in the correction terms separately and they are difficult to determine precisely, their approximate values may safely by used.

F. Kühnen and P. Furtwängler made a series of very careful and precise measurements of the absolute value at Potsdam, Germany in 1906 by this method. The value they obtained was (981.274 ± 0.003); this had long been adopted as the international standard value of gravity of the world, and gravity values in other countries had been determined by relative measurements referring to Potsdam until the accuracy of their measurements began to be questioned. The details of this circumstance will be described in the next chapter.

2.4 Falling-body method

One of the most direct ways to find the absolute value of gravity is to observe the motions of a falling body over a distance of one or two metres.

G. Atwood (1746–1807) built an apparatus which is schematically shown in Figure 2.5. A flexible string runs around a circular wheel which has a very small moment of inertia and which can rotate freely around its central axis. At both ends of the string, two masses M are suspended and they are balanced as in a pulley well. When a small mass m is put on one of the two masses, they will begin to move and their velocity v, after a time t, is given by

$$v = \frac{mt}{2M + m} g \tag{2.13}$$

or

$$g = \frac{v}{t} \frac{2M + m}{m}, \tag{2.14}$$

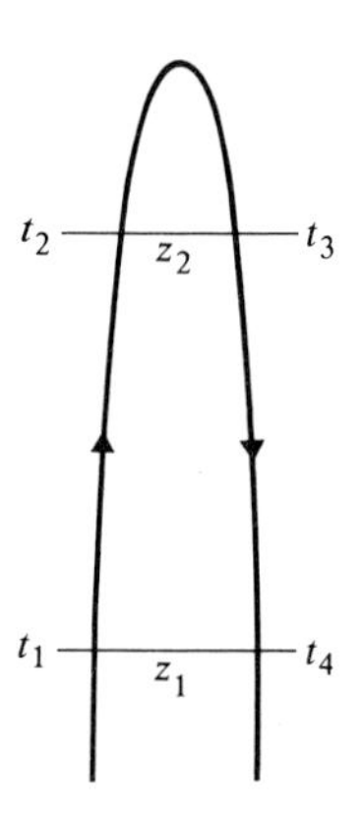

Figure 2.5 Path of a mass thrown vertically upward. t_1 and t_4 are the transit times of the mass when it crosses z_1, and t_2 and t_3 are those when it crosses z_2.

neglecting the moment of inertia of the wheel and the frictional force acting at its axis. If m is very small compared with M, the velocity v will also be very small. If the small mass m is taken away instantaneously at $t = t$, the two masses M will continue moving with this small velocity v which can easily be measured. By combining v, t, M and m, the value of g can be calculated. Results of present-day precision cannot, however, be expected from this rudimentary arrangement.

In recent years, a very accurate method for measuring the absolute value of g has been developed, first by Volet (1952) at the International Central Bureau of Weights and Measures, France. This method uses the freely falling motion of a body and is called the falling-body method.

The position z of a body which it has at $t = t$, after it is released at $t = 0$, is given by

$$z = z_0 + v_0 t + \tfrac{1}{2} g t^2, \tag{2.15}$$

where z_0 is the initial position of the body and v_0 its vertical velocity at the instant when it is released. Let z_1, z_2 and z_3 be three successive levels and t_1, t_2 and t_3 be the times of transit of the falling body across these three levels. Then

$$z_1 = z_0 + v_0 t_1 + \tfrac{1}{2} g t_1^2,$$

$$z_2 = z_0 + v_0 t_2 + \tfrac{1}{2} g t_2^2,$$

$$z_3 = z_0 + v_0 t_3 + \tfrac{1}{2} g t_3^2.$$

Eliminating z_0 and v_0 from these equations, we obtain

$$g = \frac{2\{(z_1 - z_2)(t_1 - t_3) - (z_1 - z_3)(t_1 - t_2)\}}{(t_1 - t_2)(t_1 - t_3)(t_2 - t_3)}. \tag{2.16}$$

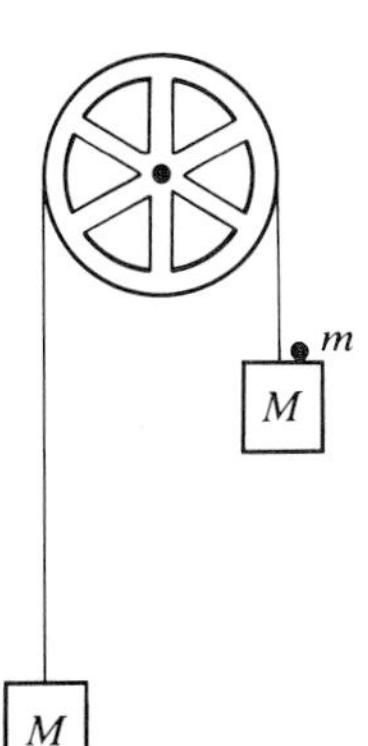

Figure 2.6 Atwood's apparatus. A flexible string goes around a wheel and two masses M are suspended at each end and balanced. An additional small mass m makes the masses move slowly.

By measuring the vertical separations $(z_1 - z_2)$ and $(z_1 - z_3)$ and the time intervals $(t_1 - t_2)$, $(t_1 - t_3)$ and $(t_2 - t_3)$, the absolute value of g at this point can be found from this relation. The time intervals can be measured very accurately by electronic devices.

A further improvement which has been introduced is to throw a body vertically upwards instead of dropping it freely. The body will pass the same level twice, when it is going up and down. Let the instants at which the body passes the two levels z_1 and z_2 be t_1 and t_2 when it is moving upwards and be t_3 and t_4 when it is moving downwards, respectively. Then

$$
\begin{aligned}
z_1 &= z_0 + v_0 t_1 - \tfrac{1}{2} g t_1^{\,2},\\
z_2 &= z_0 + v_0 t_2 - \tfrac{1}{2} g t_2^{\,2},\\
z_3 &= z_0 + v_0 t_3 - \tfrac{1}{2} g t_3^{\,2},\\
z_4 &= z_0 + v_0 t_4 - \tfrac{1}{2} g t_4^{\,2},
\end{aligned}
$$

and from these, we obtain

$$
g = \frac{8(z_2 - z_1)}{(t_4 - t_1^{\,2})^2 - (t_3 - t_2)^2}\,. \tag{2.17}
$$

In this method, the only parameters that have to be measured are the height separation $(z_1 - z_2)$ and the time intervals $(t_1 - t_4)$ and $(t_2 - t_3)$. The reported accuracy of the gravity value obtained by this method is as high as 0.006 mgal. However, it should not be forgotten that the field of gravity force through which the body moves is not perfectly uniform. The value of g at higher and lower levels of the instrument differs a little because of their difference in distance from the centre of the Earth, and the above mathematical relations do not hold exactly. If the height difference of the two levels is 1 metre, for instance, the difference in the value of g at the two levels is 0.3 mgal, which is not negligibly small in this precise measurement.

2.5 Relative measurements of gravity by means of physical pendulums

As was stated before, one of the methods for relative measurements of gravity is to swing one and the same rigid pendulum at various places and to compare its periods of oscillation at these places. Pendulums used for this purpose are mostly about 30 cm in length and have a period of about 1 second. It is important that the physical constants of the pendulums should not be affected by surrounding conditions. Effects of temperature variation or of the Earth's magnetic field should be avoided. Various non-magnetic materials have been used for this purpose, for example, brass or tungsten. The material which is most widely used now is fused silica. This material has a very small thermal-expansion coefficient (0.4×10^{-6}) and is also easily malleable. A metallic knife edge, usually of special steel, is fixed to the pendulum and rests on a perfectly flat agate seat, and the pendulum is made to oscillate around this edge in an evacuated vessel.

Usually, three pendulums of the same size are used as one set. They are swung in an evacuated vessel with their knife edges arranged to be parallel. The pendulum assembly which was constructed by the Geographical Survey Institute, Tokyo in 1951 is shown in Figure 2.7. This three-pendulum arrangement is used in order to eliminate the possible effects of ground motions on the periods of oscillation. This is necessary if the measurements are to be made on soft ground where pulsatory movements are large and disturb the motion of the pendulums. Since the three pendulums are placed on the same seat, they

Figure 2.7 A three-pendulum apparatus for relative measurements of gravity constructed at the Geographical Survey Institute, Tokyo, in 1951.

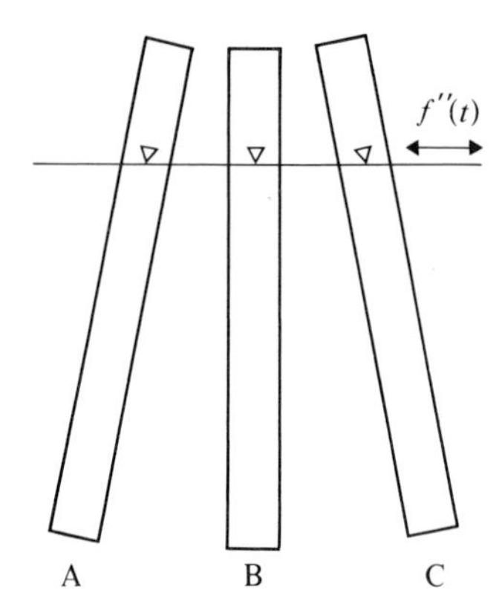

Figure 2.8 Three pendulums A, B and C placed on the same seat. The differences of oscillation between A and B, and B and C are observed in order to eliminate effects of the ground acceleration $f''(t)$.

are subjected to the same accelerations of the ground motion if there are any. If the circular frequencies of the three pendulums, A, B and C, are made to be equal to n, then the equations of motion of the pendulums subjected to ground motions $f(t)$ are

$$\ddot{x}_A + n^2 x_A = -pf''(t), \tag{2.18}$$

$$\ddot{x}_B + n^2 x_B = -pf''(t), \tag{2.19}$$

$$\ddot{x}_C + n^2 x_C = -pf''(t), \tag{2.20}$$

where $f''(t)$ is the ground accelerations and p the physical constant of the pendulums common to all of them. Subtracting one of the equations from the remaining two, we obtain

$$\ddot{\overline{x_A - x_B}} + n^2(x_A - x_B) = 0, \tag{2.21}$$

$$\ddot{\overline{x_C - x_B}} + n^2(x_C - x_B) = 0. \tag{2.22}$$

These equations mean that although x_A, x_B and x_C themselves may be disturbed by ground pulsations, the differences $(x_A - x_B)$ and $(x_C - x_B)$ will not, and will oscillate with the frequency n with which each x would oscillate when there is no disturbance. $(x_A - x_B)$ and $(x_C - x_B)$ can be observed by appropriate optical arrangements. It is usual to swing the pendulums A and C in opposite phases and the pendulum B is just hung free. This idea of taking the difference of the pendulums' motions was invented in the Netherlands where the ground is generally weak and not steady.

It is advisable to make gravity pendulums in the form of a minimum pendulum. A minimum pendulum is one that is supported by an axis at a position such that its period of oscillation is a minimum. The period of oscillation of a pendulum around an axis varies according to its distance h from the centre of gravity as was shown in Figure 2.3. The period changes according to the position of the axis and is shortest when the axis is at a certain point between A and B or between C and D. The period of oscillation of a physical pendulum is generally

$$T = 2\pi\sqrt{\frac{k^2 + h^2}{gh}}, \tag{2.23}$$

where k is the radius of gyration of the pendulum around its centre of gravity. The value h for which the period is a minimum is known from the relation

$$\frac{\partial T}{\partial h} = 0$$

and this is satisfied by

$$h = k.$$

The period is a minimum when the distance between the axis of oscillation and the centre of gravity is equal to the radius of gyration around the centre of gravity of the pendulum. The minimum pendulum has an advantage that the change in the period caused by accidental yielding of the knife edge is reduced to zero, because $\partial T/\partial h = 0$.

2.6 Measurement of oscillation period

Assuming we have a good pendulum, our next task is to measure its period of oscillation as accurately as possible. The period of oscillation of a pendulum under gravity is inversely proportional to the square root of g

$$T \propto \frac{1}{\sqrt{g}}. \tag{2.24}$$

A change in g and the corresponding change in T are related as follows:

$$\frac{\delta T}{T} = -\frac{1}{2}\frac{\delta g}{g}. \tag{2.25}$$

If we use a pendulum of which the period is about 1 second and if an accuracy of 1 mgal in g is aimed at, then

$$\frac{\delta T}{T} \approx -\frac{1}{2}\frac{10^{-3}}{980}$$

$$\approx -\frac{1}{2 \times 10^6}.$$

This means that the period has to be measured with an accuracy of one part in two million. For this purpose, an ingenious method which is called the coincidence method has long been used. Suppose the period of a pendulum is $(1 + \varepsilon)$ second, where ε is very small. If the oscillation of this pendulum is observed intermittently at intervals of exactly 1 second, the phase angle of the pendulum will be seen to shift gradually until, after a certain length of time C, these shifts will be accumulated to be 2π (= 0). This time interval is called the coincidence interval, because this is the time interval between two consecutive coincidences of the phases of the pendulum and of 1 second time signal. If the pendulum period is a little longer than 1 second, it will oscillate $(C - 1)$ times within C seconds, while if it is a little shorter, the pendulum will oscillate $(C + 1)$ times. The period of oscillation of the pendulum is therefore

$$T = \frac{C}{C \pm 1}. \tag{2.26}$$

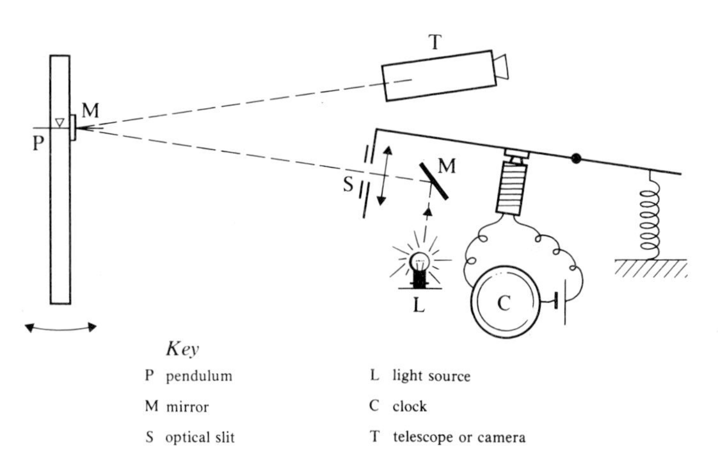

Figure 2.9 Apparatus to measure the coincidence interval. The slit is moved electromagnetically every second by the signals from the clock.

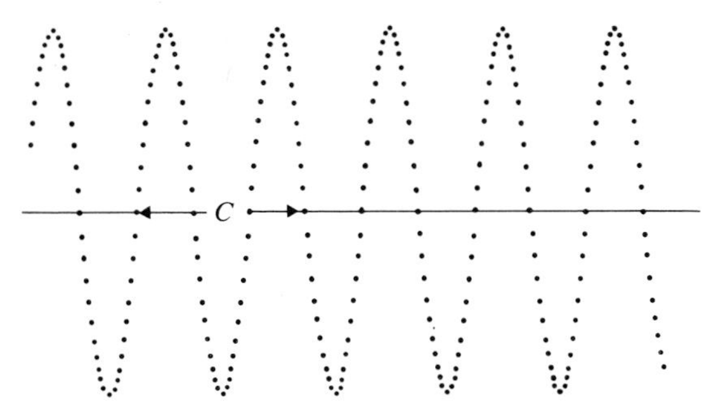

Figure 2.10 An example of a coincidence curve. The dots are marked at exactly one-second intervals. C is a coincidence interval.

From this, it follows that

$$\frac{\delta T}{T} = \pm \frac{\delta C}{C(C \pm 1)} \approx \pm \frac{\delta C}{C^2}, \tag{2.27}$$

provided C is large compared with 1. If the period of the pendulum is $(1 \pm 1/50)$ s, for instance, the coincidence interval C will be 50 s, so that

$$\frac{\delta T}{T} \approx \pm \frac{\delta C}{C^2} = \pm \frac{\delta C}{2500}.$$

If we are to achieve an accuracy of one part in two million in the period determination, the accuracy in the determination of the coincidence interval C should be

$$\delta C = 5 \times 10^{-7} \times 2500 \approx 1.3 \times 10^{-3} \text{ s}. \tag{2.28}$$

This means that if C is determined with an accuracy of one part in a thousand, the period of the pendulum can be determined with a higher accuracy of one part in two million. This useful relationship arises from the fact that $\delta T/T$ is inversely proportional to the square of C. This does not mean, however, that the longer the coincidence interval, the easier it is to find the accurate period of the pendulum. If the coincidence interval C is very long, the phase shift of the pendulum position will be very slow, and it will be more and more difficult to decide at which instant the phase difference is zero. Practically, the most appropriate length of the coincidence interval is around 50 s.

In order to find this coincidence interval an optical arrangement, as shown in Figure 2.9, is conveniently used. An optical slit is moved electromagnetically at every second according to time signals coming from a standard clock, so that a light beam goes out from it and is reflected back by a mirror fixed to the pendulum. The reflected light is observed by a telescope or is recorded on a moving photographic film of a camera. Figure 2.10 is an example of the record which was obtained in this way. The time intervals between two consecutive dots are all 1 second and the pendulum makes one oscillation between them. From this record, the coincidence interval C can be read and the period of oscillation of the pendulum calculated from it. Table 2.1 shows an example of these calculations.

Table 2.1 Calculation of period of pendulum.

No.	*Time of coincidence (h: min: s)*	*No.*	*Time of coincidence (h: min: s)*	*nC (h: min: s)*
0	12: 17: 31.9	n	13: 55: 31.4	1: 37: 58.5
1	18: 30.5	$n+1$	56: 29.0	58.5
2	19: 28.6	$n+2$	57: 26.9	58.3
3	20: 26.0	$n+3$	58: 25.2	59.2
4	21: 23.8	$n+4$	59: 22.7	58.9
5	22: 21.1	$n+5$	14: 00: 20.1	59.0
6	23: 18.7	$n+6$	01: 17.2	58.5
7	24: 15.9	$n+7$	02: 15.2	59.3
8	25: 13.9	$n+8$	03: 12.8	58.9
9	26: 11.5	$n+9$	04: 10.3	58.8

Notes: Mean of nC = 1: 37: 58.79 = 5878.79 ± 0.032 s. Approximate C determined from observations 0 and 9 = (12: 26: 11.5 − 12: 17: 32.9) ÷ 9 = 57.62 s. Approximate value of n = 5878.79 ÷ 57.62 = 102.03. Value of C = 5878.79 ÷ 102 = 57.635 ± 0.003 s. Period of the pendulum = 57.635 ÷ 56.635 = 1.0176569 ± 0.0000001 s.

The time signals from the clock used in this measurement should also have the corresponding accuracy of one part in two million. This means that the rate of the clock should be known with an accuracy of 0.04 s per day.

$$24 \times 60 \times 60 \times 5 \times 10^{-5} = 0.04 \text{ s.}$$

The rate of a clock with this accuracy can be found by comparing the clock time with wireless time signals emitted from time stations. If these are not available in remote areas, we must resort to classical astronomical observations of star positions.

Fundamentally, it is not necessary to stick to a particular unit of 1 second in relative measurements, because it is only the ratio of the oscillation periods of a pendulum at different places that comes into question and the unit of time is arbitrary so far as it is reliably constant. The present author has once used the following method.

Prepare two pendulums a and b and swing them together at a place A and measure the ratio $T_A{}^a : T_A{}^b$ by means of a coincidence apparatus operated by the signals coming from pendulum b. Then bring pendulum a to another place B where the gravity value is to be determined. Pendulum b is swung at station A as before and short wireless wave signals are sent according to its oscillations. Receiving these signals at station B, the coincidence apparatus is operated according to T_A^b and the ratio $T_B^a : T_A^b$ is measured. By comparing $T_A^a : T_A^b$ and $T_B^a : T_A^b$, the difference in the values of g at A and B can be found. In this method, a clock can be dispensed with and so there is no need to worry about the rate of the clock.

Prior to this, Bullard (1933) of Cambridge, England used the following method. He used commercial wireless telegraph signals for identifying simultaneity at two distant places. He prepared two pendulums a and b and swung them together at a point A. By comparing the number of oscillations of the two pendulums within a certain duration of time, the ratio $T_A^a : T_A^b$ could be found. Then he brought one of the pendulums, a, to another place B and made it swing there. Pendulum b was swung at A as before. By referring to the same commercial wireless telegraph signals, he could identify the simultaneity at A and B. By counting the number of oscillations of the two pendulums at distant places during the same duration of time, $T_B^a : T_A^b$ could be found. By comparing this with $T_A^a : T_A^b$, the relative gravity value at A and B could be found. If the accuracy of counting the number of oscillations of the pendulum in this method is 1/200, the duration of time t that is needed to achieve an accuracy of 1 mgal in gravity difference will be

$$t = 1/200 \div (5 \times 10^{-7}) \approx 10^4 \text{ s}$$

or a few hours. By this method, Bullard was successful in finding relative gravity values at the African Rift Valley with respect to Cambridge.

These methods of utilising wireless communications between a base station and a field station are interesting and useful, but it is inconvenient that two teams of observers, one at the base station and the other at the field station, are needed. Also, wireless signals are often disturbed by atmospheric and other electrical noises which make simultaneous reception of the signals difficult.

In recent years, quartz-crystal clocks have been remarkably improved and it is now possible to determine the period of a gravity pendulum easily by comparing it directly with crystal-clock beats. In modern apparatus, the number of oscillations of a quartz crystal within the time of n oscillations of the pendulum is counted directly by means of an electronic device, and an accuracy of 0.2 mgal in gravity measurements can be obtained.

2.7 Correction to observed period

The period of oscillation of a pendulum depends on its amplitude. If the amplitude is α radians, the period T_α is given by

$$T_\alpha = T_0(1 + \tfrac{1}{16}\alpha^2 + \cdots) \tag{2.29}$$

where T_0 is the period of the same pendulum swinging with an infinitely small amplitude. T_α increases with α. If an accuracy of 1 mgal in gravity value is aimed at, the required accuracy for the period is 5×10^{-7} s. No correction is needed for finite amplitudes if

$$\tfrac{1}{16}\alpha^2 \leq 5 \times 10^{-7}.$$

The values of α that easily satisfy this inequality are

$$\begin{aligned} \alpha &\leq \sqrt{(16 \times 5 \times 10^{-7})} \\ &\approx 3 \times 10^{-3} \\ &= 10'. \end{aligned}$$

But if α is larger than this, which is usually the case, corrections should be made on T_α in order to find T_0. Since the correction should be accurate up to 5×10^{-7},

$$\begin{aligned} \delta(\tfrac{1}{16}\alpha^2) &= \frac{\alpha}{8}\,\delta\alpha \\ &= 5 \times 10^{-7}. \end{aligned}$$

$\delta\alpha$, which is required for the amplitude measurement, depends on α itself and is given by

$$\delta\alpha \leq \frac{4 \times 10^{-6}}{\alpha}.$$

This relation shows that the larger the amplitude α, the more accurately it should be measured. The allowable limit of $\delta\alpha$ calculated for various α are given in Table 2.2.

Table 2.2 Allowable limits of $\delta\alpha$.

α	$\delta\alpha$
10′	3′.3
20′	2.2
30′	1.5
40′	1.1
50′	0.87
1°	0.74
2°	0.37
3°	0.25
4°	0.19
5°	0.14

Since the accuracy of amplitude measurements is about 1′, the amplitude of oscillation itself should be smaller than 1°.

In the above, the knife edge, around which the pendulum oscillates, has been regarded as an infinitely thin straight line. If this is actually the case, the pressure acting on it due to the weight of the pendulum is infinitely large and the result must be that the knife edge is pressed down to form a minute rolling cylinder, as shown in Figure 2.11 with some exaggeration. Some experiments have shown that T_α for various amplitudes does not necessarily change according to the theoretical relation

$$T_\alpha = T_0(1 + \tfrac{1}{16}\alpha^2). \tag{2.30}$$

This disagreement is probably due to the rolling of the knife edge.

The knife edge of the pendulum should be kept accurately horizontal when its period is to be measured. If the knife edge is inclined to the

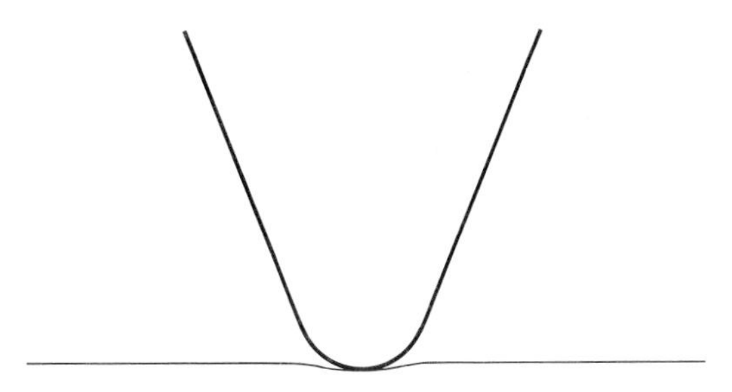

Figure 2.11 An exaggerated sketch of deformations of the knife edge of a pendulum and the seat on which it rests.

horizontal direction by an angle i, $g \cos i$ (instead of g) makes the pendulum oscillate and the period T_i is then given by

$$T_i = \frac{T_0}{\sqrt{\cos i}} . \tag{2.31}$$

If i is small

$$\sqrt{\cos i} = \left(1 - \frac{i^2}{2}\right)^{1/2}$$

$$= 1 - \frac{i^2}{4} .$$

In order to guarantee an accuracy of 5×10^{-7} in the period determination, $i^2/4$ must be smaller than 5×10^{-7}, whence

$$i \leq 1.4 \times 10^{-3} \approx 5'.$$

The knife edge of the pendulum should be kept horizontal with an accuracy of $5'$.

2.8 Holweck–Lejay's inverted pendulum

In its ordinary method of measurement by means of a physical pendulum, gravity is the direct motive force which makes the pendulum oscillate. This is the reason why its period of oscillation should be measured with a high accuracy. If, on the other hand, this main motive force of gravity could be balanced by some other force, then the change in gravity will produce a much larger change in the period of oscillation of the pendulum. Using this idea, F. Holweck and R. P. Lejay of France in 1930 constructed an inverted pendulum which is schematically shown in Figure 2.12. By means of an elastic blade spring fixed to its bottom, the pendulum is kept upright. When the pendulum deviates from the vertical by an angle θ, an elastic moment $\varepsilon\theta$ acts from the blade to the pendulum in the direction to make it stand upright, while a moment due to gravity $mgl\theta$ will act on it to make it fall. If $\varepsilon > mgl$, then the pendulum is stable and oscillates with a period

$$T = 2\pi\sqrt{\frac{I}{\varepsilon - mgl}} \tag{2.32}$$

about its neutral point, where I is the moment of inertia of the pendulum

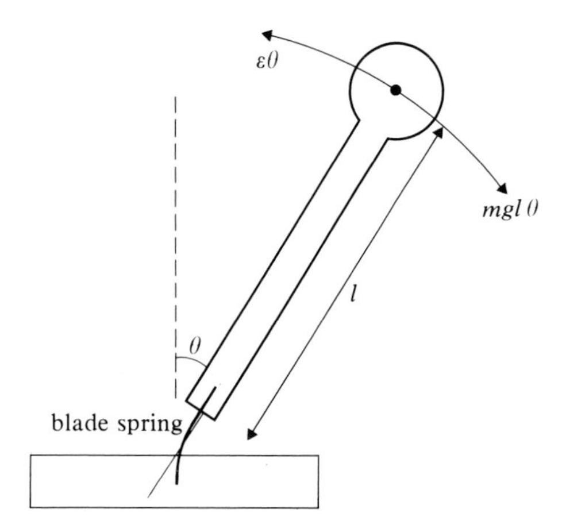

Figure 2.12 Holweck–Lejay's pendulum (1930). $\varepsilon\theta$ is the moment caused by the elastic blade and $mgl\theta$ is that caused by gravity.

around 0, l the distance between its centre of gravity and 0, and m its mass. If $(\varepsilon - mgl)$ could be made small, T will be large. This period varies with g. If a change in g is δg, the change δT in the period is given by

$$\frac{\delta T}{T} = \frac{1}{2} \frac{ml}{\varepsilon - mgl} \delta g. \tag{2.33}$$

The period of oscillation T' when this pendulum is swung freely as an ordinary pendulum is

$$T' = 2\pi \sqrt{\frac{I}{mgl}} \,. \tag{2.34}$$

In this case, the change in T' corresponding to a change δg is

$$\frac{\delta T'}{T'} = -\frac{1}{2} \frac{\delta g}{g} \,. \tag{2.35}$$

The changes in periods for the two cases are related as follows:

$$\frac{\delta T/T}{\delta T'/T'} = -\frac{mgl}{\varepsilon - mgl} = -\left(\frac{T}{T'}\right)^2 . \tag{2.36}$$

If T could be made to be five times as long as T', for instance, then $\delta T/T$ would be 5^2, which is 25 times as large as $\delta T'/T'$, and this would make measurements much easier.

Unfortunately, things do not take place simply as they are expected. One of the largest difficulties is that the elastic constant ε of the blade spring at the bottom of the pendulum changes sensitively with temperature. Since $(\varepsilon - mgl)$ is already very small, even a small variation in ε causes a large percentage variation of $(\varepsilon - mgl)$ and hence the period T. Because of these drawbacks, the Holweck–Lejay pendulum has not come to be widely used.

2.9 Gravimeters

The methods of measurement of gravity by means of a pendulum are called dynamical methods, because its dynamical property is utilized. In contrast to these, there are other methods in which no part of the instrument is moving. These methods are called static methods. The simplest arrangement is to use an elastic spring from which a mass is suspended as shown in Figure 2.13. If the length of the spring is l_0 when no mass is suspended, and l when a mass m is suspended, the spring is in equilibrium when the elastic force produced by its elongation is equal to the gravity force acting on the mass

$$mg = \varepsilon(l - l_0) \tag{2.37}$$

where ε is the elastic constant of the spring. At another place, where the gravity value is $(g + \delta g)$, the elongation of the spring will change according to

$$\frac{\delta(l - l_0)}{l - l_0} = \frac{\delta g}{g}. \tag{2.38}$$

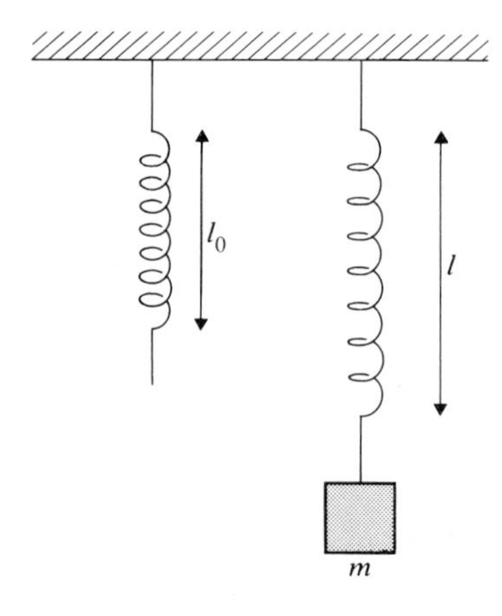

Figure 2.13 Elongation of an elastic spring caused by the weight of a mass m.

From the change in elongation $\delta(l - l_0)$, the change in gravity can be found. If δg is 1 mgal

$$\frac{\delta(l - l_0)}{l - l_0} = \frac{\delta g}{g} \approx 10^{-6}.$$

Even for a very weak spring for which $l - l_0 = 1$ m, the change in its elongation for 1 mgal is only 1 micron (10^{-3} mm). In order to make the change in elongation larger $(l - l_0)$ itself should be made larger. To make the change in elongation 0.1 mm for a 1 mgal change in gravity, for instance, $(l - l_0)$ should be as long as 100 m. This is practically out of the question.

The period T of up and down oscillations of a helical vertical spring carrying a mass m is

$$T = 2\pi\sqrt{\frac{m}{\varepsilon}}$$

$$= 2\pi\sqrt{\frac{l - l_0}{g}}. \tag{2.39}$$

This is equal to the period of free oscillation of a simple pendulum having a length $(l - l_0)$. To make a pendulum with large $(l - l_0)$ is to make a vertical pendulum with a long period of oscillation. This has also been a problem in seismometry for designing a long-period vertical seismograph. For this purpose, several arrangements have been devised. The simplest of these is shown in Figure 2.14. In this arrangement, the period of oscillation is $\sqrt{(a/b)}$ times as long as that when the mass is directly suspended from the spring. The gravity force acting at the lower end of the spring may be regarded as having been reduced to $(b/a)g$ according to its geometrical configuration. If b/a could be made very small, the period could be made very long. Unfortunately, however, this pendulum can be stable only within a very narrow range of

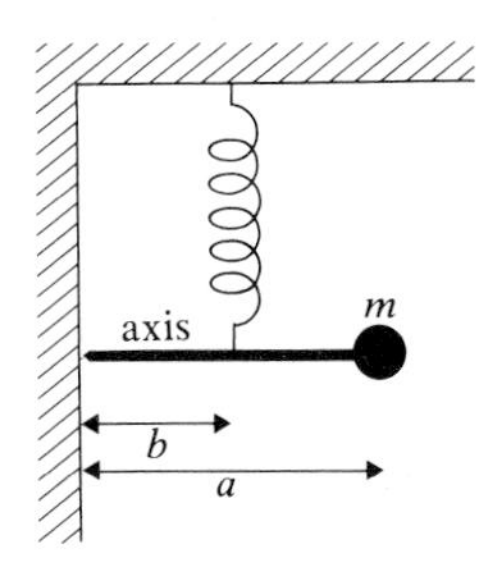

Figure 2.14 A rod suspended horizontally by an elastic spring.

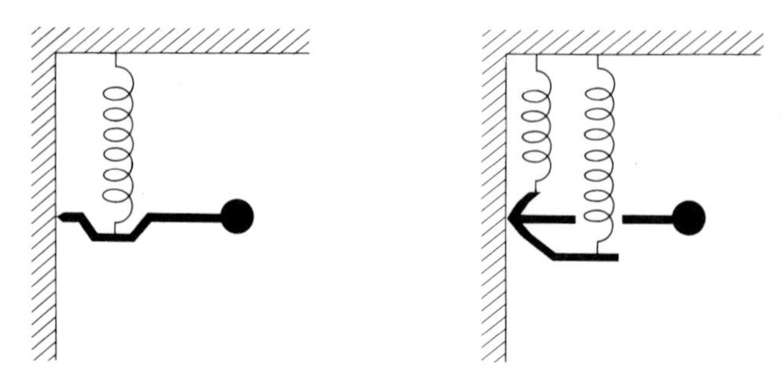

Figure 2.15 Various methods for suspending rods horizontally by elastic springs.

deflection. Several other arrangements which are shown in Figure 2.15 have been devised in order to get rid of these inconveniences.

A much more ingenious arrangement was invented by LaCoste (1934) of France. Referring to Figure 2.16 the clockwise moment M_s due to the gravity force acting on the pendulum is

$$M_s = mga \sin \theta .$$

For the spring, a wire is wound in one plane in the form of a spiral, like the so-called 'hair-spring' used in a watch. This spring is called a zero-length spring because its length is zero when it does not carry a mass. When the end of the spring is moved from 0 to r, the clockwise moment M_ε due to the elastic force of the spring is

$$M_\varepsilon = -\varepsilon rs . \tag{2.40}$$

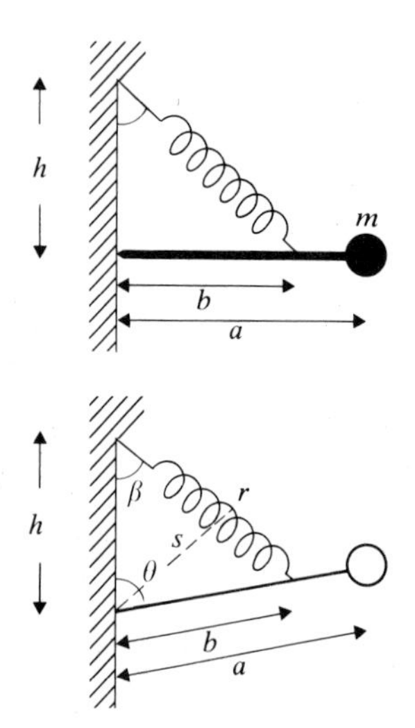

Figure 2.16 LaCoste's zero-length spring pendulum (1957). By choosing instrumental constants appropriately, the periods of oscillation can be made very long.

But, from geometry, since

$$\frac{r}{\sin\theta} = \frac{b}{\sin\beta},$$

$$s = h\sin\beta,$$

it follows that

$$M_s = -\varepsilon\frac{b\sin\theta}{\sin\beta}h\sin\beta \tag{2.41}$$

$$= -\varepsilon bh\sin\theta.$$

The total moment M acting on the pendulum is therefore

$$M = M_s + M_\varepsilon = (mga - \varepsilon bh)\sin\theta. \tag{2.42}$$

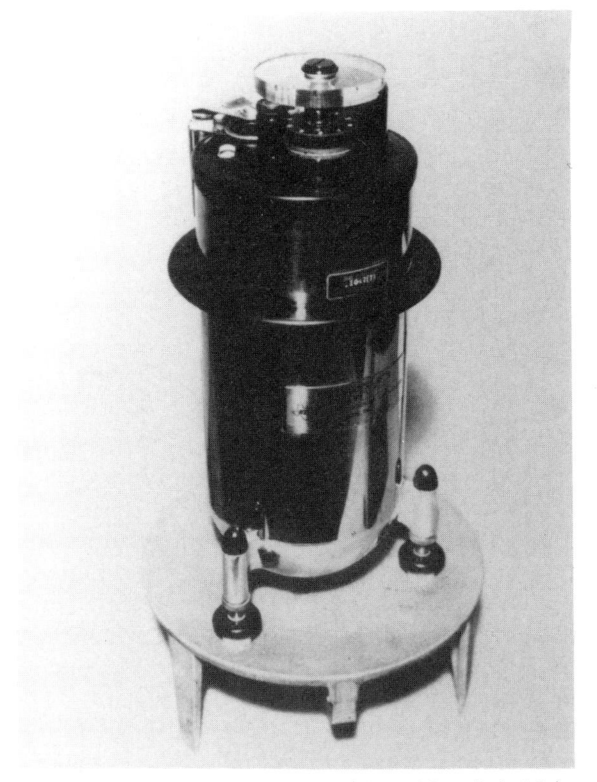

Figure 2.17 Worden's gravimeter (about 30 cm in height).

If, by choosing the instrumental constants appropriately, $(mga - \varepsilon bh)$ could be made to be very small, the period of oscillation of this pendulum would be very long. By this arrangement, a very long period can be obtained with a wide range of deflection. If this spring is used, the equilibrium position will react sensitively to a small change in gravity and this can be used as a good static instrument for relative measurements of gravity.

There are of course many difficulties to solve technically before a good gravimeter can be made: how to keep the elastic constant and dimensions of the spring unchanged notwithstanding alterations in surrounding temperature; how to avoid the effect of motion of the surrounding air; how to avoid the effect of 'drift' of the spring according to time; how to keep the air pressure constant inside the pendulum box; how to make everything compact and easy to carry, and so on. Various types of gravimeters have been built since the 1940s. The gravimeters known by the names of Worden, LaCoste and Romberg, North-American are some of these, which are on the market. They are only a few kilograms in weight. The obtainable accuracy is 0.01 mgal and the time needed for a measurement at one place is 5–10 minutes. Figure 2.17 is a photograph of a Worden gravimeter.

However carefully these gravimeters are designed and constructed they cannot be entirely free from the effects of drift of the spring used. Also the elastic constant of the spring cannot be expected to be strictly constant with time. In order to reduce these uncertainties in gravity surveys, it is advisable to reoccupy the same point as frequently as possible for estimating the amount of drift, and to calibrate the sensitivity of the gravimeters by occasionally going to those points where gravity differences have been accurately measured by standard pendulum measurements.

2.10 Early statical measurements

Instruments which can be regarded as prototypes of modern gravimeters were constructed earlier. Naturally, these are not useful for present-day precision, and descriptions of these old instruments will have little practical significance apart from historical interest.

Other than those already mentioned, it should be noted that several attempts were made earlier to build instruments without using an elastic spring. Hecker (1908) of Germany tried to find the value of gravity by measuring the boiling point of pure water. The boiling point of water is the temperature at which its vapour pressure is equal to the atmospheric pressure P working from outside, and therefore it varies according to the pressure. By measuring the boiling temperature accurately, the absolute value of the outside atmospheric pressure P can be found theoretically, in dyne cm^{-2}. On the other hand, P is related to the height h of the mercury column of a mercury barometer by the relation

$$P = \rho g h \tag{2.43}$$

where ρ is the density of mercury. Combining the values of P known from the measurements of the boiling point of water and the height h of the mercury column of the barometer, the value of g can be calculated by means of this relation. Experimentally, however, this method is too difficult to be successful, because of the high accuracy needed in measuring the boiling point and the height of the mercury column.

Haalck (1931) of Germany constructed a gas gravimeter as shown in Figure 2.18. The difference in pressure p_1 and p_2 in the two air chambers is balanced by the height h of the mercury column, so that

$$\rho g h = p_2 - p_1. \tag{2.44}$$

If g changes, h will also change. By measuring the change in h, the change in g can be found. In order to overcome the difficulty in measuring the change in h which is very small, Haalck put a light liquid called toluene ($C_6H_5CH_3$) on the mercury surface and the liquid is led through very thin tubes connected to the mercury vessels. If the mercury surfaces are Q in area and the narrow tubes are q in area, then a small change in h will be magnified Q/q times in the change in position of the meniscus of toluene in the thin tubes. If the radius of the mercury

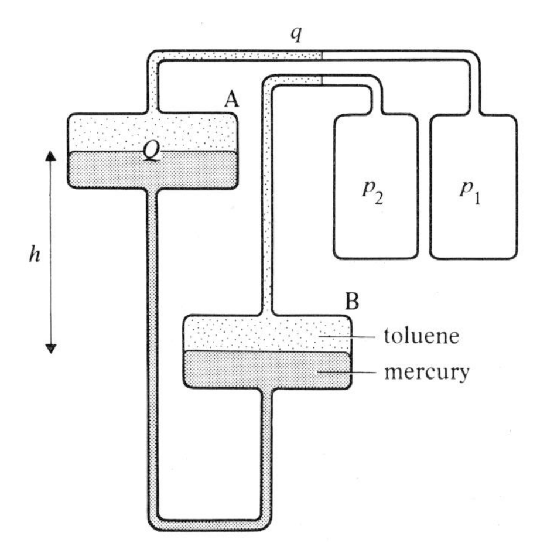

Figure 2.18 Haalck's gas gravimeter. p_1, p_2 are air pressures, Q is the cross-sectional area of the mercury surfaces, and q the cross-sectional area of the toluene capillaries.

surface is 10 cm and that of the narrow tubes is 1 mm, for instance, then

$$Q/q = 10^4$$

and a small change in h will be much easier to measure. There are several difficulties in accomplishing this for the instrument to be of practical use. One of them is how to keep the temperature of the air inside the chambers constant and another is to avoid the cohesion of toluene liquid to the walls of the thin tubes. Both of these have a considerable influence on the results of the measurements.

If actual substances behave exactly as described in an idealistic formula, simple static gravimeters can be made to be useful for accurate gravity measurements. A buoy gravimeter, which is shown in Figure 2.19, is one of these. A buoy containing air and a mercury column is made to float on water. The enclosed air is compressed by the weight of the mercury column. A thin rod which is attached to the buoy projects from the surface of the water and the floating equilibrium is maintained.

At a place where g is a little larger, the mercury surface will be lowered and the whole volume of the buoy decreases, so that the buoy will sink until it takes a new position of equilibrium. If the rod attached to the buoy is very thin, the lowering of the buoy will be large enough to be measured. At a place where g is smaller, the buoy will rise. In this way, a small change in g will cause a sensible change in the equilibrium position of the buoy. In actual fact, the motion of the surrounding water, the effects of the contact angle of the thin rod and water, and of contamination of the water surface, etc. make it almost impossible to use this buoy for the purpose of gravity measurement.

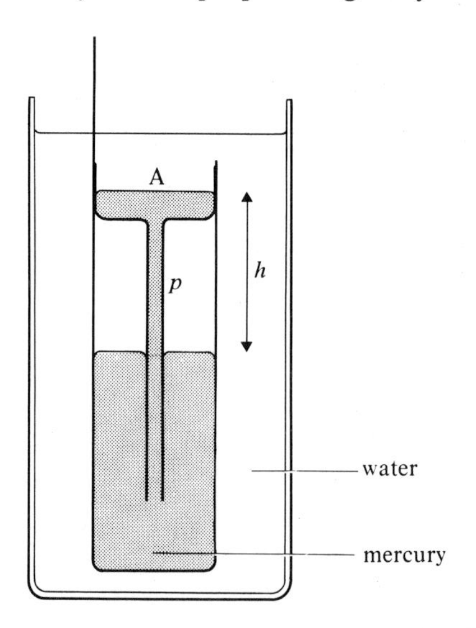

Figure 2.19 Buoy gravimeter, where p is the air pressure.

2.11 Gravity measurements aboard a submarine

Gravity measurements on seas are very desirable in view of their large area but these involve many difficult problems which are hard to overcome. The most fundamental point is, of course, that no stable platform is available for setting up a gravity instrument. Vening Meinesz (1929) of the Netherlands was the first to be successful in overcoming the difficulty and he measured gravity values at sea with an accuracy of a few mgal. He noted that a submarine sails very quietly under water and will provide a steady platform on which to place a gravity instrument.

Although accelerations in a submarine are small, they are not so small as to be ignored. In order to eliminate the effects of acceleration of the vessel, Vening Meinesz adopted an arrangement using three gravity pendulums A, B and C in the following way. The three pendulums are made to be equal in size and are swung on the same seat in the same direction; the differences of oscillation of A, B and C are observed. Oscillation of each of the pendulums is disturbed by the motion of the floor on which the pendulum case is placed, but because the three pendulums are placed on the same seat, they are disturbed equally, so that the differences (A − B) and (B − C) will not be disturbed. These differences are recorded on moving photographic paper in a camera fixed to the pendulum case.

Another important problem is how to keep the knife edges of the pendulums horizontal with the required accuracy, notwithstanding the movement of the floor. As was already mentioned before, the knife edges of the pendulums must be kept horizontal to an angle accuracy of 5′ in order to obtain an accuracy of 1 mgal in gravity value. In order to satisfy this requirement, Vening Meinesz suspended the pendulum box from a frame fixed to the floor. He used a gimbal suspension which consists of a pair of axes at right angles to each other, as shown in Figure 2.20, so that the box is free to tilt in any direction. If the floor tilts, the knife edges of the gravity pendulums will keep their horizontal directions by means of this gimbal suspension. In order to avoid accidental oscillations of the pendulum box, blades are attached to its

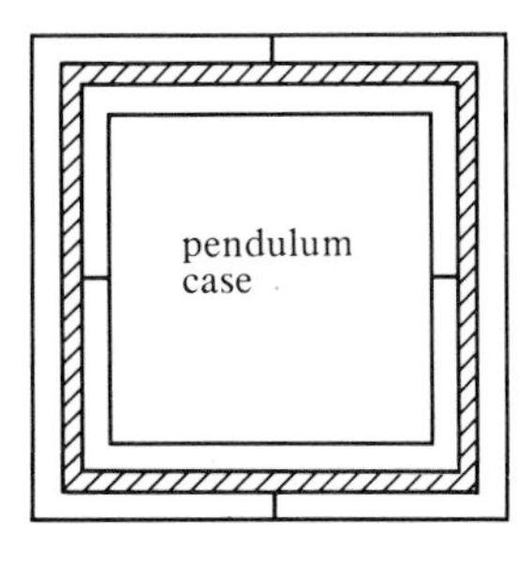

Figure 2.20 A gimbal suspension. The pendulum case can be tilted in any direction by the actions of the two axes.

bottom and they are put in a tank (not shown in Figure 2.21) which contains a viscous oil. Figure 2.21 shows the whole apparatus as it is installed in a submarine.

Vening Meinesz made extensive voyages across the Indian, Pacific and Atlantic oceans on board a submarine of the Royal Dutch Navy, and made gravity measurements at several hundred points. His results, especially the gravity distribution along the archipelago of the East Indies, are one of the most exciting discoveries in geophysics in this century and exerted a great influence on subsequent research on the structure of the Earth's crust.

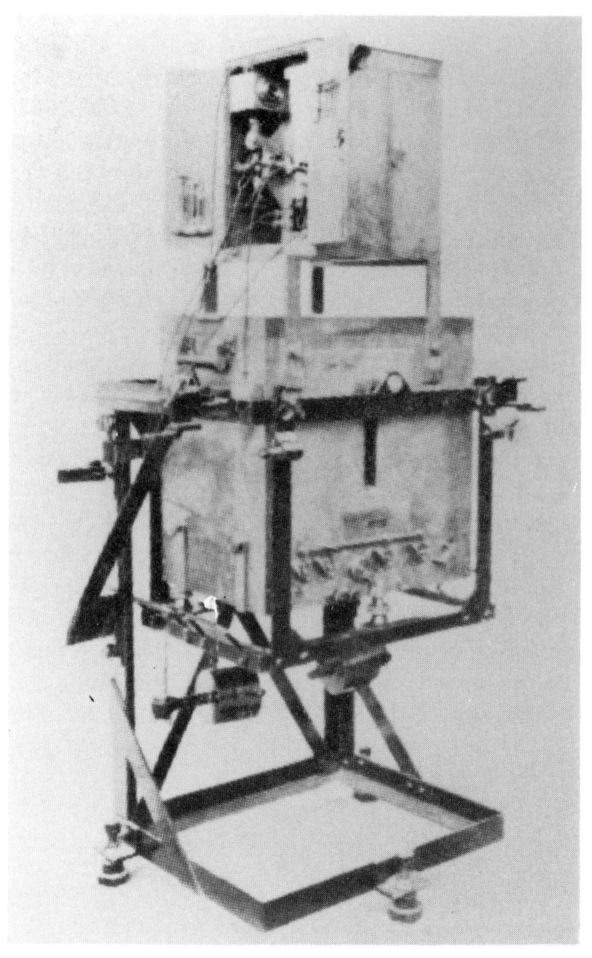

Figure 2.21 Vening Meinesz's apparatus (1929). The upper case contains the photographic recording camera, the middle case the three-pendulum case, and the black frame the gimbals.

The pendulum arrangement of Vening Meinesz, however, has some features which require further consideration. If the floor of the submarine on which it is installed becomes inclined, an undesirable occurrence will take place. Although the pendulum box will tend to keep its horizontal direction owing to the action of the gimbal suspension, the oil damper will drag it to produce an angular deflection. Also, the motion of the submarine floor may involve horizontal displacements. These horizontal motions will induce movements of the pendulum box which are more or less like a seismometer pendulum. The period of free oscillation of the pendulum box being several seconds, the deflection of the box caused by the motion of the floor may easily exceed $5'$ in inclination. There are two ways of overcoming these dynamical difficulties of Vening Meinesz's apparatus. One is to lengthen the period of oscillation of the pendulum box in order to reduce its seismometer-like motions, and the other is to improve the damping mechanism in order to avoid the undesirable viscous coupling of the pendulum box and the submarine floor.

If the horizontal motion of the floor is a in amplitude, T its period, and the period of the pendulum box is T_0, then the inclination angle α of the pendulum box caused by this motion is

$$\alpha = \frac{4\pi^2 a}{gT_0^2} \frac{1}{1 - (T/T_0)^2} . \tag{2.45}$$

If a is 100 cm and T is 5–6 s, then T_0 should be 50–60 s in order that α will be smaller than $5'$. In order to make such a long-period suspension system, the moment of inertia I of the pendulum box should be very large and the distance h between its centre of gravity and the rotation axis very small, since

$$T_0 = 2\pi\sqrt{\frac{I}{Mgh}} . \tag{2.46}$$

With a large value for I and a small value for h, the box system will be something like a balancing toy. Also, in order to avoid the undesirable coupling between the pendulum box and the floor, a hollow circular tube is fixed to the box system with a viscous liquid in it. Damping will be effected by the relative motion of the tube wall and the liquid contained in it. If the viscosity of the liquid is too great the tube and the liquid will move together and no damping will result. If the viscosity of the liquid is too small, no damping will result. A certain relationship must be satisfied between the period of the pendulum box, the viscosity of the liquid and the cross-sectional width of the tube in order to obtain

Figure 2.22 Tsuboi's pendulum apparatus (1934). The pendulum case is supported like a balancing toy in order to make its period long; circular tubes are filled with oil and work as dampers.

the largest possible damping. The present author (Tsuboi 1934) built an apparatus according to this principle, which is shown in Figure 2.22.

This idea of using a long-period balancing pendulum was later adopted by Vening Meinesz himself. He placed a small long-period balancing pendulum side-by-side with the main gravity pendulum apparatus, observed the direction differences of the two, and made the necessary corrections.

2.12 Browne's remark

In connection with gravity measurements on a floor which is moving up and down, Browne (1937) of Cambridge, England made the following important remark. He pointed out that the gravity value calculated from the average period of a gravity pendulum on a moving floor is not the average of gravity values.

If $z''(t)$ is the time-to-time vertical acceleration of the floor, the time-to-time frequency n of a gravity pendulum oscillating on this floor is

$$n(t) = \frac{1}{2\pi\sqrt{l}} \sqrt{\{g + z''(t)\}}, \tag{2.47}$$

where l is the length of the equivalent pendulum. Vertical acceleration on the deck of a ship is around 100 gal or $g/10$ in order of magnitude, so that the above expression of $n(t)$ can be expanded in a power series of z''/g which is about $\frac{1}{10}$.

$$n(t) = \frac{1}{2\pi}\sqrt{\frac{g}{l}}\left\{1 + \frac{z''(t)}{g}\right\}^{1/2}$$

$$\approx \frac{1}{2\pi}\sqrt{\frac{g}{l}}\left\{1 + \frac{1}{2}\frac{z''}{g} - \frac{1}{8}\left(\frac{z''}{g}\right)^2\right\}.$$

Taking the average value $\overline{n(t)}$, the average of z'' will vanish, but the average of $(z'')^2$ will not vanish. Thus

$$\overline{n(t)} = \frac{1}{2\pi}\sqrt{\frac{g}{l}}\left\{1 - \frac{1}{8}\overline{\left(\frac{z''}{g}\right)^2}\right\}$$

$$= \frac{1}{2\pi}\sqrt{\frac{g}{l}}\left\{1 - \frac{1}{4}\overline{\left(\frac{z''}{g}\right)^2}\right\}^{1/2}. \tag{2.48}$$

This means that the apparent value of g calculated from the average period of the pendulum becomes smaller by a factor $1 - \frac{1}{4}(z''/g)^2$. Even if the vertical acceleration of a floor is as small as 10 gal,

$$1 - \frac{1}{4}\left(\frac{z''}{g}\right)^2 = 0.99998$$

and the apparent decrease of the gravity value will be as much as 20 mgal. In order to be able to make corrections for this effect, continuous measurements of the vertical acceleration of the floor are needed.

2.13 Gravity measurements on surface ships

Although Vening Meinesz's method using a submarine has proved successful, it is not always easy to use a submarine for this purpose. Attempts have naturally been made to develop methods for measuring gravity values on surface ships. There have been two different directions of approach for this purpose. One is to use static gravimeters and the other is to use string gravimeters.

In using static gravimeters for this purpose, the following three points should be considered fully.

(a) How to keep the gravimeters vertical the whole time.
(b) How to eliminate the effect of the vertical acceleration of the ship.
(c) How to record or observe the equilibrium positions of the gravimeters.

The first problem can be solved by putting the gravimeter on a gyroscope or on a platform which is stabilized by a gyroscope. Since the gravimeter is not very heavy, this can be done without great technical difficulties. As to the second problem, it should be noted that the period of oscillation of a ship is 5–10 s and its vertical acceleration is of the

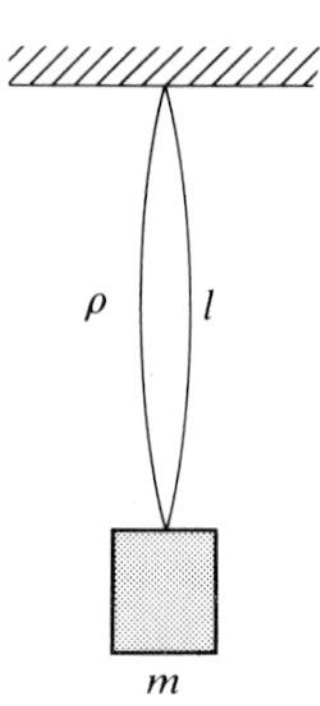

Figure 2.23 A chord-vibration gravity meter.

order of 100 gal and that, in contrast to this, variation of gravity from one place to another in a voyage is much slower in time and smaller in magnitude. In order to measure variation in gravity with an accuracy of 1 mgal, the change in the output of the gravimeter due to the motion of the ship must be reduced by a factor of 10^{-5}. This means that the output of the gravimeter should be passed through a low-pass and high-cut filter which has the required characteristics.

An instrument based on another principle was constructed by Y. Tomoda of Tokyo in 1963 and this is called the TSSG (Tokyo Surface Ship Gravity Meter). This instrument uses the vibrations of a string. As is shown in Figure 2.23, when a string of length l which carries a mass m at its lower end is made to vibrate as a chord, its frequency f is

$$f = \frac{1}{2l}\sqrt{\frac{mg}{\sigma}}, \tag{2.49}$$

where σ is the linear density of the string. Change in f and change in g from one place to another are related as follows:

$$\frac{\delta f}{f} = \frac{1}{2}\frac{\delta g}{g}. \tag{2.50}$$

Tomoda used a beryllium–copper ribbon, 25 mm in length, 0.1 mm in width, and 0.01 mm in thickness. The suspended mass is 20 g. The frequency of vibration of the ribbon is about 1500 per second. The ribbon is placed in a magnetic field and the vibration of the ribbon is maintained automatically by an electric current, which is produced through it by its vibration in the magnetic field and fed back to it. The motion of the suspended mass caused by the motion of the ship is prevented by means of four thin wires stretched horizontally between it and the wall of the instrument. The whole vibrator is small in size and is mounted on a gyroscope to keep it vertical.

Figure 2.24 Tomoda's gravity meter (TSSG) using chord vibrations of a string. The whole assembly is mounted on a gyroscope.

Vibrations of the ribbon take place according to the sum of gravity and vertical acceleration of the ship. The time needed for 850 vibrations is measured by a quartz–crystal oscillator and when 1000 such measurements (about 500 s) are made, the values are put into an electronic filter designed for this purpose, high-frequency components are cut out and the gravity value corresponding to the central instant of 500 s is calculated by an electronic computer and printed. The ship's position, east–west velocity and the depth of the sea are also automatically measured and printed. Tomoda and his collaborators have already made several voyages extending over one million km. Figure 2.24 shows this instrument.

Another type of string-vibration gravity meter has been built at the Massachusetts Institute of Technology (MIT), USA. In this instrument, two strings are used, as shown in Figure 2.25. They are stretched to a tension T. The tension acting in the upper ribbon is $(T + mg)$, while that acting in the lower ribbon is $(T - mg)$. If f_1 and f_2 are the frequencies of the upper and lower ribbons respectively,

$$f_1 = \frac{1}{2l\sqrt{\sigma}}\{T + mg(t)\}^{1/2}$$

$$= \frac{\sqrt{T}}{2l\sqrt{\sigma}}\left[1 + \frac{1}{2}\frac{mg(t)}{T} - \frac{1}{8}\left\{\frac{mg(t)}{T}\right\}^2\right] \qquad (2.51)$$

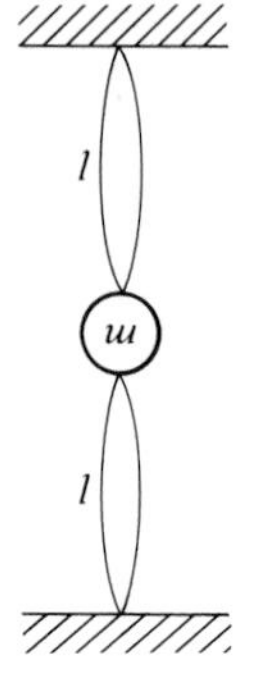

Figure 2.25 Two-vibration chord gravity meter developed at MIT.

$$f_2 = \frac{1}{2l\sqrt{\sigma}}\{T - mg(t)\}^{1/2}$$

$$= \frac{\sqrt{T}}{2l\sqrt{\sigma}}\left[1 - \frac{1}{2}\frac{mg(t)}{T} - \frac{1}{8}\left\{\frac{mg(t)}{T}\right\}^2\right] \tag{2.52}$$

then

$$f_1 - f_2 = \frac{1}{2l}\cdot\frac{mg(t)}{T}\sqrt{\frac{T}{\sigma}}, \tag{2.53}$$

and from this difference, $g(t)$ can be found. This difference changes slowly according to $g(t)$ and changes rapidly according to the ship's motion. Here again, a low-pass and high-cut filter are needed as in the case of Tomoda's instrument.

References and further reading

Browne, B. C. 1937. Second order corrections to pendulum observations. *Month. Not. R. Astron. Soc., Geophys. Suppl.* **4**, 271.

Bullard, E. C. 1933. The observation of gravity by means of invariable pendulums. *Proc. R. Soc., Lond.* **141**, 233.

Bullard, E. C. 1936. Gravity survey measurements in East Africa. *Phil Trans R. Soc., Lond.* **235**, 445.

Gilbert, R. L. C. 1949. A dynamic gravimeter of novel design. *Proc. Phys. Soc., Lond.* **62**, 445.

Haalck, 1931. Ein statischer Schwerkraftmesser. *Z. Geophys.* **7**, 95.

Haalck, H. 1932. Ein statischer Schweremesser. *Z. Geophys.* **8**, 17, 197.

Haalck, H. 1933a. Berichte über den genenwärtigen Stand der Entwickelung des statischen Schweremessers. *Z. Geophys.* **9**, 81.

Haalck, H. 1933b. Neue Messungsergebnisse mit dem statischen Schweremesser. *Z. Geophys.* **9**, 285.

Haalck, H. 1936. Der neue statische Schweremesser des Geodätischen Instituts in Potsdam. *Z. Geophys.* **12**, 1.

Hecker, O. 1903. *Bestimmung der Schwerkraft auf den Atlantischen Ozean sowie in Rio de Janeiro*. Veröff. Preuss, Geod. Inst., no. 11.

Hecker, O. 1908. *Bestimmung der Schwerkraft auf dem Indischen Ozean und deren Küsten*. Ibid., no. 16.

Kater, H. 1818. An account of experiments for determining the length of the pendulum vibrating seconds in the latitude of London. *Phil Trans R. Soc., Lond.* **108**, 32.

LaCoste, L. J. B. 1934. A new type long period vertical seismograph. *Physics* **5**, 178.

LaCoste, L. J. B. and H. N. Carkson 1957. Improvements in tidal gravity meters. *Trans Am. Geophys. Union* **38**.

Tomoda, Y. 1961. Tokyo surface ship gravity meter (in Japanese). *J. Geol Soc., Japan* **7**, 116.

Tomoda, Y., J. Segawa and T. Takemura 1972. Comparison measurements of gravity at sea using a TSSG and a Graf-Askania sea gravimeter. *J. Phys Earth* **20**, 267.

Tomoda, Y., C. Tsuboi and H. Kanamori 1961. Continuous measurements of gravity on board a moving surface ship. *Proc. Jap. Acad.* **37**, 571.

Tsuboi, C. 1933. On the possibility of taking gravity measurements at sea on board an ordinary vessel. *Proc. Imp. Acad., Japan* **9**, 374.

Tsuboi, C. 1934. Improved apparatus for taking gravity measurements at sea on board an ordinary vessel. *Proc. Imp. Acad., Japan* **10**, 640.

Tsuboi, C. and T. Fuchida 1937. Relative measurements of gravity by means of short wireless wave communications (in Japanese). *J. Seis. Soc., Japan* **9**, 546.

Vening Meinesz, F. A. 1929, *Theory and practice of gravity measurements at sea.* Delft: Waltman. (2nd edn 1941, Neth. Geod. Comm.)

Volet, Ch. 1952. Mesure de l'acceleration due à la pesanteur, au Pavillon de Breteuil. *C.R. Acad. Sci, Paris* **235**, 442.

Wing, C. G. 1967. An experimental deep sea-bottom gravimeter. *J. Geophys. Res.* **72**, 1249.

3 Absolute values, reduction, standard formula and gravity anomalies

3.1 Absolute value of gravity

Gravity values have been and still are measured at a great many points on the Earth. In order that these values be useful for global geophysical studies, they should not be merely accumulations of numerical values but should be referred to the same standard absolute value and be interrelated and coordinated to form a unified system. For several decades, the absolute value of gravity – determined at the Geodätisches Institut at Potsdam, Germany, by means of a reversible pendulum by Kühnen and Furtwängler (1906) – was adopted as the international standard and gravity values at all other points in the world were directly or indirectly tied to Potsdam by relative measurements. The absolute value of gravity at Potsdam found at that time was

$$g_P = 981.274 \pm 0.003.$$

The worldwide network system based on this value is called the Potsdam system.

Thirty years after the Potsdam measurement, however, its accuracy began to be doubted. This problem came to notice when Heyl and Cook (1936) made an absolute measurement at Washington, DC, USA and Clark (1939) made another at Teddington, England. Both of these results were found to be not quite compatible with the Potsdam value. The absolute values of gravity at Washington and Teddington were found to be

$$g_W = 980.080$$

and

$$g_T = 981.182$$

respectively.

The differences of g_P, g_W and g_T had been known to be

$$g_W - g_P = -1.174 \text{ gal},$$

$$g_T - g_P = -0.079 \text{ gal},$$

$$g_T - g_W = 1.097 \text{ gal},$$

by relative measurements. Using the values of g_W, $(g_W - g_T)$, g_T and $(g_T - g_P)$, the value g_P was deduced as follows:

$$g_P = g_W + (g_P - g_W) = 980.080 + 1.174 = 981.254,$$

$$g_P = g_T + (g_P - g_T) = 981.182 + 0.079 = 981.261.$$

Both of these values are significantly smaller, by 10 mgal on the average, than the authorized value determined at Potsdam to be 979.274 before. On the other hand,

$$g_T - g_W = 1.102 \text{ gal}$$

by the absolute measurements and

$$g_T - g_W = 1.097 \text{ gal}$$

by the relative measurement, and these two differences agree within 5 mgal. Although this agreement cannot be said to be very satisfactory, the discrepancy between the observed g_P and the calculated g_P is even larger. So there were good reasons to suspect that the old Potsdam value was in error. It was thought, however, that no hasty steps should be taken to change the standard value at Potsdam until more data had been accumulated which would justify the revision.

Not long after this, extremely careful absolute measurements were made at 10 points around the world with accuracies better than 1 mgal. Among them are Bagoda, Columbia, a high altitude point near the Equator, and Fairbanks, Alaska, USA, a lowland point at a latitude greater than 60° N. The former is a representative point of low gravity, while the latter that of high gravity. Combining the results of the absolute measurements made at the 10 points and the relative-gravity differences between each of the points and Potsdam, the gravity value at Potsdam was deduced. All the results agreed in showing that the old Potsdam value was too large.

A number of geophysicists from various countries met at a conference to discuss this problem. They agreed that a vast mass of calculations for establishing a self-consistent international network of gravity values should be made, taking into account not only the results of absolute measurements at the 10 points but also 1200 relative measurements by means of gravity pendulums and about 24 000 relative measurements by means of gravimeters, by which a number of points in the world have been interconnected gravimetrically. The final results of the enormous calculations were reported at the General Assembly of the International Union of Geodesy and Geophysics held at Moscow in 1971, and were adopted. According to this, the gravity value deduced for Potsdam was

$$g_P = 981.26019 \pm 0.000017$$

which is 14 mgal smaller than the old value.

The old Potsdam measurement had been believed to be a model of the highest precision and accuracy, every conceivable source of error having been investigated, eliminated and corrected. It was hardly understandable that it contained such a large error as 14 mgal. Dryden (1942), of the US Bureau of Standards, critically reviewed the old Potsdam report and could not find anything to be revised in it, except that a correction used in the old report for the effect of elastic deformation of the reversible pendulum on its period was not quite adequate. When this correction term was modified properly, it was found that the old value decreased by 15 mgal. This meant the end of the Potsdam system.

By way of the worldwide adjustment calculations, the gravity values at 1854 points in the world could also be determined on a unified system. This worldwide net is called the International Gravity Standardization Net 1971 (IGSN 71). In most countries, there are at least several points that are included in this number. All precise gravity measurements which are to be made in a country should refer to these points.

In Japan, for instance, 39 points are included in IGSN 71. H. Suzuki of the Geographical Survey Institute, Tokyo, thought in 1975 that this number was not large enough for a country in which the gravity distribution is very complicated. He added 83 more points, which had been already connected to some of the 39 points by relative measurements, and made net adjustment calculations. Thus there are now 122 points in Japan which are evenly distributed geographically in the country and are internationally and domestically interconnected to form a rigid network with an accuracy better than 1 mgal. The geographical distribution of these 122 points is shown in Figure 3.1.

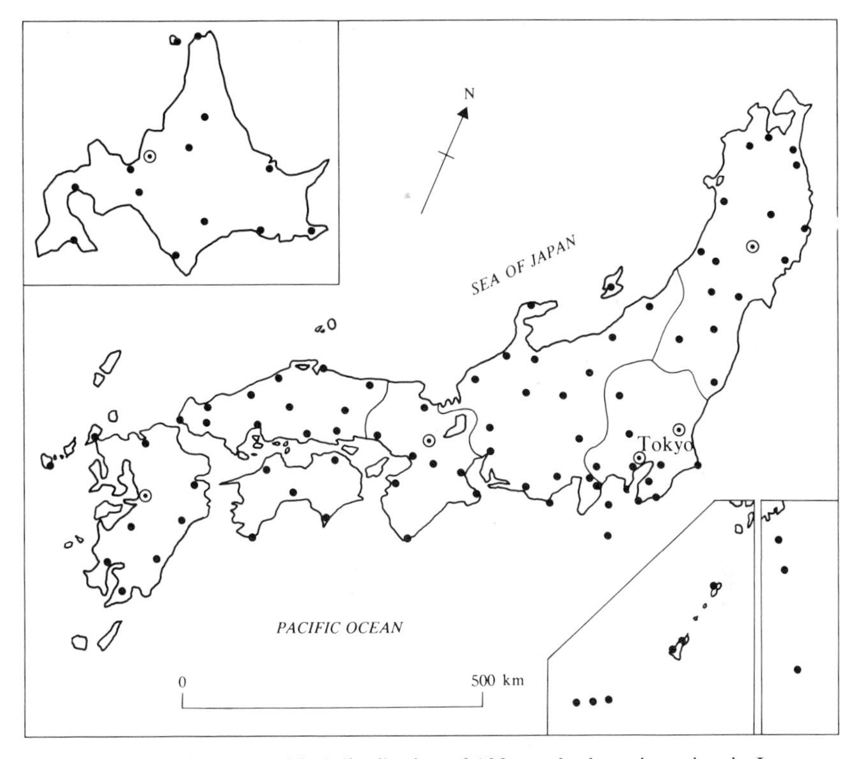

Figure 3.1 The geographical distribution of 122 standard gravity points in Japan.

3.2 Reductions of observed gravity values

The points at which gravity values have been measured in the world are at different heights and are situated in various topographical environments: high mountains, flat lands and the ocean. The gravity values at these places are affected by attractions of surrounding topographical masses. In order that these values be used for studying the worldwide distribution of gravity or for investigating their distribution in a certain area, they have to be reduced to the values that would be observed in the case of a standardized topographical feature. The process of doing this is called **gravity reduction**.

The first thing to be considered is the difference in height of the points. The attraction of the Earth at a point decreases with its height according to the inverse square of its distance from the centre of the Earth. If the radius of the Earth is R and the height above sea level of a gravity point is h, the value of gravity, g, and that at sea level right under it, g_0, are related as follows:

$$g = g_0 \frac{R}{(R+h)^2}\,. \tag{3.1}$$

From this, it follows that

$$g_0 = g\left(1 + \frac{h}{R}\right)^2 \approx g + \frac{2g}{R}h.$$

To find the value of g_0 from the observed g, $(2g/R)h$ must be added to g. Numerically, for h in metres

$$\frac{2g}{R}h \approx 0.3086 \times h \text{ mgal}$$

that is, 1 mgal must be added to the observed value for each 3.3 m elevation. This reduction for height is called the free-air reduction. For example, at the summit of Mt Fuji (3776 m), the free-air reduction amounts to

$$0.3086 \times 3776 = 1165 \text{ mgal} = 1.165 \text{ gal}.$$

This numerical value of the vertical gradient of gravity has been derived for the average Earth. The actual value of the gradient at each point may differ from this, because of the effect of the attraction of anomalous mass beneath the point. This problem will be discussed later.

The next thing to be considered in gravity reductions is the effect of attraction of the rock mass near the gravity points. We have to calculate the downward component of attraction due to the rock mass and subtract it from the observed gravity values. For instance, the downward component of attraction T at the summit of an isolated circular conical mountain due to its rock mass, as shown in Figure 3.2, is

$$\begin{aligned} T &= 2\pi G\rho \int_0^a \frac{rh}{(r^2+h^2)^{3/2}}\,\mathrm{d}r \\ &= 2\pi G\rho h\left\{1 - \frac{h}{\sqrt{(a^2+h^2)}}\right\}. \end{aligned} \tag{3.2}$$

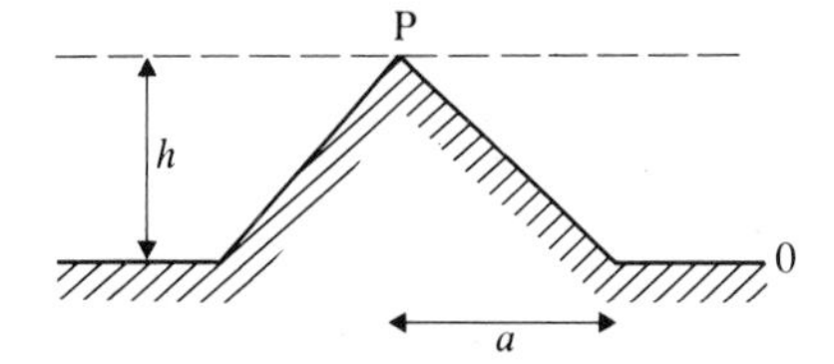

Figure 3.2 An isolated circular conical mountain: h = height; a = radius of the base.

If $h = 4000$ m, $a = 10$ km, and $\rho = 2.67$, the downward component of attraction T at the summit is

$$T = 2\pi \times 6.67 \times 10^{-8} \times 2.67 \times 4 \times 10^{5}\left\{1 - \frac{4}{\sqrt{(4^2 + 10^2)}}\right\}$$

$$= 285 \text{ mgal.} \tag{3.3}$$

Even if an isolated mountain is not as big as this, the reduction for its mass may exceed 100 mgal. But usually topographies are much gentler and the reduction for their masses is much smaller.

The correction for topographical masses is usually carried out in the following two steps. Consider a horizontal plane that passes a gravity point as shown in Figure 3.3. The actual topographical surface is higher

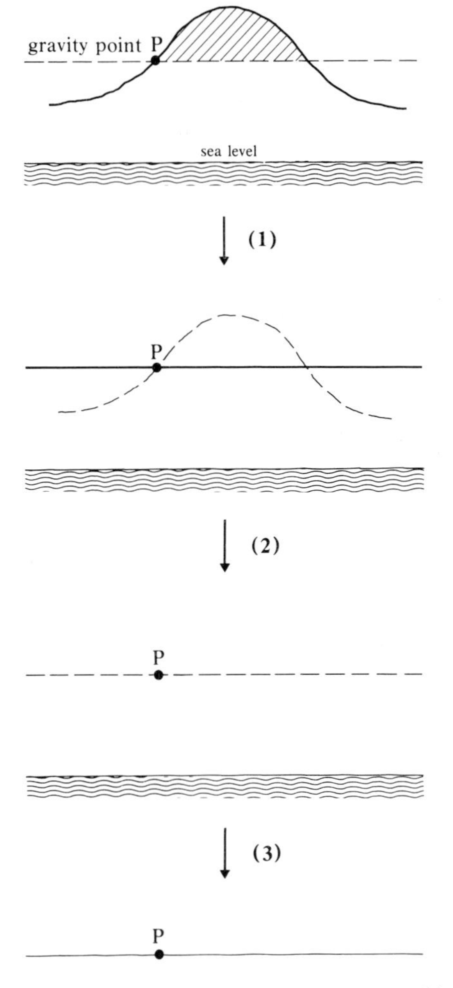

Figure 3.3 Three steps for gravity reduction. (1) Reduction to the horizontal terrain. (2) Bouguer reduction. (3) Free-air reduction.

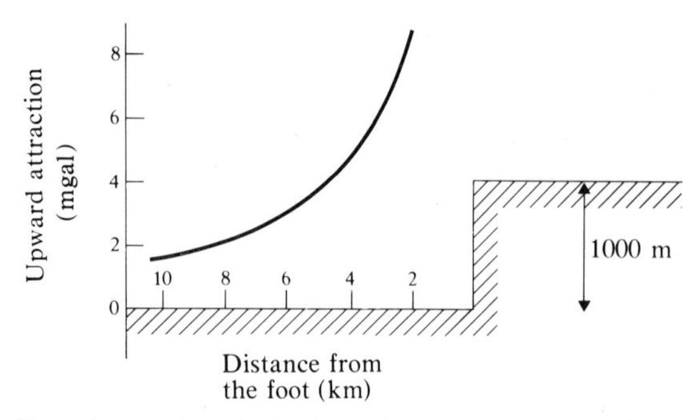

Figure 3.4 Upward attraction of a horizontal plateau at various distances from its foot.

than the plane at some places and lower at others. The attraction due to those parts higher than the horizontal plane, which is shown by shading in Figure 3.3, has an upward component. Those parts lower than the horizontal plane, which are left blank in the figure, have no mass and therefore exert no attraction. The effects of these two parts are to make the observed gravity smaller compared with the attraction of the horizontal mass of a uniform thickness h. By calculating the effects of both and taking proper account of them in the observed gravity values, the result will give the gravity value when the topography has a uniform height. This is called the reduction to the horizontal terrain. The amount of this reduction depends of course on the actual topography around the point, but in most cases it is less than 10 mgal.

Suppose, for instance, that there is a very rough topography as shown in Figure 3.4. The vertical component of attraction due to its mass at various distances from the foot of the cliff is shown by a curve in the figure. Unless the point is very close to the cliff, the reduction is smaller than 10 mgal.

Any topography can be regarded as an assemblage of many rock columns, as shown in Figure 3.5. The vertical component t of attraction due to the mass of a rock column is

$$t = G\,\rho\int_0^h \frac{rz\ \mathrm{d}\theta\ \mathrm{d}r\ \mathrm{d}z}{(r^2 + z^2)^{3/2}}$$

$$= \frac{2G\rho\ \mathrm{d}\theta\ \mathrm{d}r}{r}\left\{1 - \frac{r}{\sqrt{(r^2 + h^2)}}\right\}. \tag{3.4}$$

In order to do this calculation, a grid like that shown in Figure 3.5b is useful. Draw this grid on a transparent sheet of paper or glass and place it on a topographical map of the area with the origin at the gravity point, and read the average height in each compartment of the grid. After

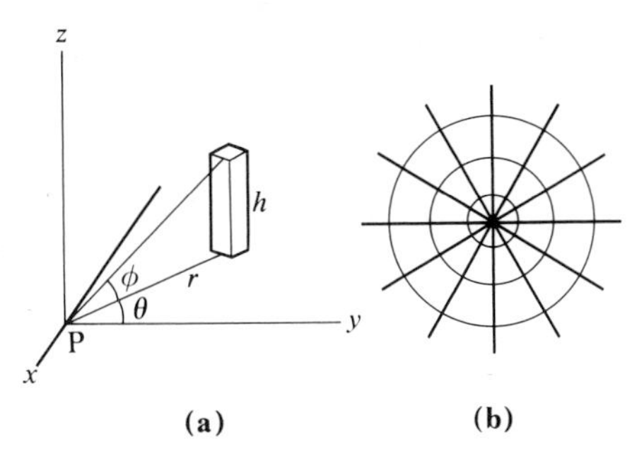

Figure 3.5 (a) Vertical component of attraction due to the mass of a rock column. (b) A mesh useful for calculations.

reading r and h for each compartment, calculate the values of t according to the above formula and add them together for all the compartments. But seeing that

$$\frac{1}{r}\left\{1-\frac{r}{\sqrt{(r^2+h^2)}}\right\}$$

tends to zero rapidly with r, this reduction to the horizontal terrain may be safely omitted in most cases. The reason for this rapid decrease of t with r is that the factor

$$\left\{1-\frac{r}{\sqrt{(r^2+h^2)}}\right\} \approx -\frac{1}{2}\left(\frac{h}{r}\right)^2 \tag{3.5}$$

decreases as $1/r^2$ and its vertical component, which is its effect on gravity, is $1/r$ times the attraction. As a whole, the effect of a distant mass on gravity decreases as $1/r^3$.

Finally, the effect of attraction of a horizontal rock layer with a uniform thickness h must be considered. The downward attraction T of the layer is

$$T = 2\pi\rho\, Gh \tag{3.6}$$

and this must be subtracted from the observed gravity value. If $\rho = 2.67$ and h is measured in metres

$$T = 0.112\, h \text{ mgal}$$

that is, 1.12 mgal for each 10 m.

The reduction for this layer of a uniform thickness is called the **Bouguer reduction**. Pierre Bouguer (1698–1758) was a French scientist who was sent to South America to make geodetic surveys. It is said that he was impressed by the gigantic features of the Andes and he began to study the attraction of mountain masses.

As to the gravity values measured on the surface of the sea, the height of the point is zero so that no free-air reduction is needed. For making Bouguer reductions we note that, in oceanic areas, there is sea water ($\rho = 1.03$) down to a depth d instead of rock ($\rho = 2.67$), so that the relative defect of attraction, for d in metres, is

$$2\pi G(2.76 - 1.03)\,d = 0.069 \times d \text{ mgal}$$

that is, 69 mgal for 1000 m depth. This should be added to the observed value in order to reduce it to the case when there is a rock mass instead of sea water.

We have seen how to apply various reductions to the observed value of gravity. It is customary to denote the observed value of gravity by g, g + (free-air reduction) by g_0, g_0 + (reduction to the horizontal terrain and the Bouguer reduction) by g_0''.

As an example g at a point ($\rho = 35°\ 0'$ N, $h = 600$ m) is 979.598 and

$$g_0 = 979.598 + 0.195 = 979.783,$$

$$g_0'' = 979.738 - 0.067 = 979.716.$$

3.3 Standard gravity formula

Thousands of properly reduced gravity values are known over a wide range of latitudes. Our next task is to find a numerical formula which gives these values as a function of their latitudes. On a reasonable assumption that the Earth is very close to a rotational ellipsoid in form, this expression will be a function of latitude only. The mathematical form of this expression is

$$\gamma = g_e(1 + \beta \sin^2 \phi - \beta' \sin^2 2\phi), \tag{3.7}$$

where g_e is the gravity value on the Equator,

$$\beta = \frac{5}{2}\frac{\omega^2 a}{ge} - \varepsilon - \frac{17}{14}\frac{\omega^2 a}{g_e}\varepsilon,$$

$$\beta' = \frac{\varepsilon}{8}\left(\frac{5\omega^2 a}{g_e} - \varepsilon\right), \tag{3.8}$$

ω is the angular velocity of the rotation of the Earth; a is the equatorial radius of the Earth; b the polar radius of the Earth; and $\varepsilon = (a - b)/a$ is the ellipticity of the Earth. This formula will be derived in Chapter 4.

Since the gravity values are known at thousands of points with various latitudes, the numerical constants in this expression can be determined by the method of least squares. The formula based on the IGSN71 is

$$\gamma = 978.03185(1 + 0.0053024 \sin^2 \phi - 0.0000059 \sin^2 2\phi) \quad (3.9)$$

and this γ is called the standard or normal gravity. The normal gravity values at different latitudes according to this formula are given in Table 3.1.

Table 3.1 Standard gravity values at various latitudes.

Latitude	γ	*Latitude*	γ
0°	978.03185	50°	981.06948
5°	978.07107	55°	981.50655
10°	978.18755	60°	981.91695
15°	978.37780	65°	982.28813
20°	978.63611	70°	982.60872
25°	978.95472	75°	982.86890
30°	979.32402	80°	983.06068
35°	979.73289	85°	983.17816
40°	980.16897	90°	983.21773
45°	980.61905		

The ellipticity of the Earth, derived from this formula, is

$$\varepsilon = 1/298.257.$$

According to this standard formula, the difference in gravity corresponding to a difference of 1 minute of angle in latitude is as given in Table 3.2.

Table 3.2 Gravity difference corresponding to 1 minute difference in latitude.

Latitude	*mgal*
0°	0.00
10°	0.52
20°	0.97
30°	1.31
40°	1.49
50°	1.49
60°	1.31
70°	0.97
80°	0.52
90°	0.00

These values tell us that the latitude of a gravity point must be known down to 0'.07 (126 m) if an accuracy of 0.1 mgal is required in the comparison of the calculated and observed values of gravity near $\phi = 40° \sim 50°$.

3.4 Gravity anomalies

The value of gravity at a point calculated by the standard formula and that observed and deduced do not usually agree with each other. This is because the effect of attraction of an invisible anamalous mass under the point is involved in the observed value. The difference $(g_{obs} - \gamma)$ is called the **gravity anomaly**. In particular, $(g_0 - \gamma)$ is called the **free-air anomaly** and is denoted by Δg_0, while $(g_0'' - \gamma)$ is called the **Bouguer anomaly** and is denoted by $\Delta g_0''$. In the example given in the preceding section,

$$\Delta g_0 = 979.783 - 979.733 = 50 \text{ mgal},$$

$$\Delta g_0'' = 979.716 - 979.733 = -17 \text{ mgal}.$$

Gravity anomalies are smaller than ± 100 mgal at most places and it is exceptional that they exceed ± 200 mgal. Very roughly speaking, free-air anomalies are smaller than Bouguer anomalies. Bouguer anomalies on land are mostly negative, and those on oceans are mostly positive. These facts can be explained by the theory of isostasy.

The geographical distribution of gravity anomalies in an area can be shown on a map by drawing contour lines connecting points of equal anomalies. Free-air anomalies give information about the actual gravity field along the surface of the Earth; Bouguer anomalies give information about hidden underground masses.

References and further reading

Clark, J. S. 1939. An absolute determination of the acceleration due to gravity. *Phil. Trans R. Soc., Lond.* **238**, 65.

Cook, A. H. 1952. Comparison of the acceleration due to gravity at the National Physical Laboratory, Teddington, the Bureau International des Poids et Mesure, Savre, the Physikalische-technische Bundesanstalt, Brunswick, and the Geodetic Institute, Potsdam. *Proc. R. Soc., Lond.* **213**, 408.

Dryden, H. L. 1942. A re-examination of the Potsdam absolute determination of gravity. *J. Res. Nat. Bur. Stand.* **29**, 303.

Heyl, P. R. and G. S. Cook 1936. The value of gravity at Washington. *J. Res. Nat. Bur. Stand.* **17**, 505.

Kühnen, F. and Ph. Furtwängler 1906. *Bestimmung der absoluten Grösse der Schwerkraft zu Potsdam.* Veröff. König. Preuss. Geod. Inst., no. 27.

Preston-Thomas, H., L. G. Turnbull, E. Green, T. M. Dauchinee and S. N. Kalra 1960. An absolute measurement of the acceleration due to gravity at Ottawa. *Can. J. Phys.* **38**, 824.

Wollard, G. 1963. *An evaluation of the Potsdam Datum.* Sci. Rep. Hawaii Inst. Geophys., no. 1.

4 Gravity and the ellipticity of the Earth

4.1 Newton's calculation

If we wish to know the shape of the Earth as a planet, no direct and objective observation has been possible, because we have been unable to leave its surface. One of the indirect but practicable methods for making this possible was to see the distribution of gravity values over the whole surface of the Earth. The geoid, of which we wish to know the shape, is the surface of equal gravity potential. In a general sense, the problem is therefore to find the shape of a surface of equal potential from the distribution of its normal gradient on it. It should be mentioned that the recent technical developments of artificial satellite tracking have provided an entirely new approach to the problem.

The possibility of determining the shape of the Earth from gravity values on it was first noticed about 300 years ago, when a French astronomer, J. Richer (1630–96), went to Cayenne in South America ($\varphi = 4° 46'$N) to make astronomical observations. He took a pendulum clock with him which had kept accurate time in Paris. He found that this same clock lost 148 seconds per day at Cayenne. He thought that the reason for this was because the gravity value at Cayenne was smaller than that at Paris due to the elliptic shape of the Earth and that the clock swung with a longer period there than at Paris. The clock's daily loss of 148 seconds can be produced by a decrease of 3.7 gal in gravity.

Newton (1642–1727) was interested in Richer's discovery and he tried to calculate the ellipticity of the Earth from the clock data. For the calculations, he made a simple assumption that the Earth is a self-gravitating ellipsoid of revolution of a liquid with a uniform density ρ, rotating around its axis and in hydrostatic equilibrium. He imagined that two wells were dug down to the Earth's centre, one from the pole and the other from the Equator. If the two wells are connected at the bottom of each, as shown in Figure 4.1, the liquid columns in the two wells must

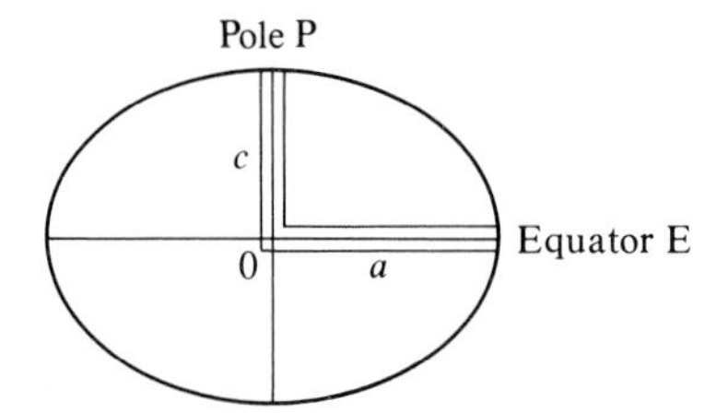

Figure 4.1 Two wells dug from the pole and from the equator to the centre of the Earth and connected.

be in balance with each other. The attraction X of the Earth at the Equator was calculated to be

$$X = \frac{2\pi\rho G(1 + \lambda^2)}{\lambda^3} a\left(\tan^{-1}\lambda - \frac{\lambda}{1 + \lambda^2}\right) \tag{4.1}$$

and that at the pole Z to be

$$Z = \frac{2\pi\rho G(1 + \lambda^2)}{\lambda^3} 2c(\lambda - \tan^{-1}\lambda), \tag{4.2}$$

where λ is the eccentricity of the Earth,

$$\lambda = \frac{\sqrt{(a^2 - c^2)}}{a}, \tag{4.3}$$

and a and c are the equatorial and polar radii respectively. The ratio of Z and X is

$$\frac{Z}{X} = \frac{2c}{a} \frac{\lambda - \tan^{-1}\lambda}{\tan^{-1}\lambda - \lambda/(1 + \lambda^2)}.$$

But since λ is very small compared with 1, the right-hand side of the above expression can be expanded in a power series of λ with the result

$$\frac{Z}{X} = 1 + \frac{1}{10}\lambda^2.$$

λ^2 is twice the ellipticity ε, because

$$\lambda^2 = \frac{a^2 - c^2}{a^2} = \frac{(a + c)(a - c)}{a^2} \approx 2\frac{a - c}{a} = 2\varepsilon.$$

Using this ε, the ratio of Z and X is

$$\frac{Z}{X} = 1 + \frac{\varepsilon}{5}.$$

The gravity g_e at the Equator is equal to the attraction X minus the centrifugal force. The centrifugal force at the Equator is

$$\omega^2 R = 3.4$$

and is about 1/289 of X. Therefore

$$g_e = X(1 - 1/289).$$

The gravity at the pole g_p is the attraction itself. Therefore

$$\frac{g_p}{g_e} = \frac{Z}{X(1 - 1/289)} = \frac{1 + \varepsilon/5}{1 - 1/289}. \tag{4.4}$$

Dividing the Earth's ellipsoid into a number of conformal ellipsoidal shells, the above relationship holds for every one of them. The ratio of hydrostatic pressures at the bottom of the two wells is then equal to

$$\text{(ratio of gravity)} \times \text{(ratio of radii)} = \frac{1 + \varepsilon/5}{1 - 1/289}(1 - \varepsilon).$$

This must be equal to 1 in order that the liquid columns in the wells will be in balance. Solving the equation

$$\frac{1 + \varepsilon/5}{1 - 1/289}(1 - \varepsilon) = 1 \tag{4.5}$$

for ε, it is found that

$$\tfrac{4}{5}\varepsilon = 1/289$$

or

$$\varepsilon = 1/231.$$

This means that the ellipticity of an ellipsoid of revolution with a uniform density, which is rotating with the same angular velocity as the Earth, can be in equilibrium only when its ellipticity is 1/231. In this case

$$\frac{g_p}{g_e} = \frac{1 + \varepsilon/5}{1 - 1/289} = \frac{1 + \varepsilon/5}{1 - \frac{4}{5}\varepsilon} \approx 1 + \varepsilon \tag{4.6}$$

and

$$\frac{g_p - g_e}{g_e} = \varepsilon = \frac{a - c}{a}. \tag{4.7}$$

The percentage difference of the polar and equatorial values of gravity is equal to that of the polar and equatorial radii. Conversely, if the ratio of gravity values at the Equator and at the pole is known, the ellipticity of the Earth can be found at once.

The proof that a self-gravitating ellipsoid of revolution which is rotating around its axis can be in equilibrium under a certain condition is omitted here.

4.2 Huygens' calculation

In contrast to Newton, Christian Huygens (1629–95) assumed an Earth in which its total mass M is concentrated at the centre. In this case

$$g_p \approx \frac{GM}{c^2} \tag{4.8}$$

$$g_e \approx \frac{GM}{a^2} - \omega^2 a. \tag{4.9}$$

A unit mass at the pole has the gravity potential GM/c and that at the Equator $(GM/a + \omega^2 a^2/2)$. The two must be equal in order that the geoid is an equipotential surface. Hence

$$\frac{GM}{c} = \frac{GM}{a} + \frac{\omega^2 a^2}{2}.$$

Putting the relation

$$c = a(1 - \varepsilon)$$

in this expression, we obtain

$$\frac{GM}{a}(1 + \varepsilon) = \frac{GM}{a} + \frac{1}{2}\omega^2 a^2,$$

whence

$$\varepsilon = \frac{1}{2}\frac{\omega^2 a}{GM/a^2} = \frac{1}{2}\frac{1}{289} = \frac{1}{578}.$$

The ellipticity of the geoid for an Earth of concentrated mass must be 1/578.

As has been seen, an Earth of a uniform density has an ellipticity of 1/231, while an Earth of concentrated mass has an ellipticity of 1/578.

The ellipticity of the Earth known at present is 1/298, which is between the above two extreme values. This shows that the density distribution within the actual Earth is also somewhere between the two extreme assumptions of Newton and Huygens. The density within the Earth increases toward its centre.

4.3 General theory – theorem of Clairaut

In the calculations of Newton and Huygens above, the density distribution within the Earth was assumed from the outset. Now let us discuss the problem generally without assuming any particular density distribution. The only thing to be assumed is that the geoid is a figure of revolution with a small ellipticity.

Using notation as in Figure 4.2,

$$e^2 = \rho^2 + r^2 - 2\rho r \cos\gamma, \tag{4.10}$$

the gravity potential W at P′ (x, y, z) is

$$W = G\int \frac{\mathrm{d}m}{e} + \frac{1}{2}(x^2 + y^2)\omega^2. \tag{4.11}$$

But

$$\frac{1}{e} = \frac{1}{r}\left\{1 - \frac{2\rho}{r}\cos\gamma + \left(\frac{\rho}{r}\right)^2\right\}^{1/2}$$

and expanding the right-hand side of this equation in a power series of ρ/r, this is reduced to

$$\frac{1}{e} = \frac{1}{r}\left\{1 + \left(\frac{\rho}{r}\right)P_1 + \left(\frac{\rho}{r}\right)^2 P_2 + \ldots\right\}$$

where

$$P_1 = \cos\gamma,$$
$$P_2 = \tfrac{3}{2}\cos^2\gamma - \tfrac{1}{2}.$$

Then the expression for W becomes

$$W = \frac{G}{r}\left(\int \mathrm{d}m + \frac{1}{r}\int P_1\rho\,\mathrm{d}m + \frac{1}{r^2}\int P_2\rho^2\,\mathrm{d}m + \ldots\right) + \tfrac{1}{2}(x^2 + y^2)\omega^2. \tag{4.12}$$

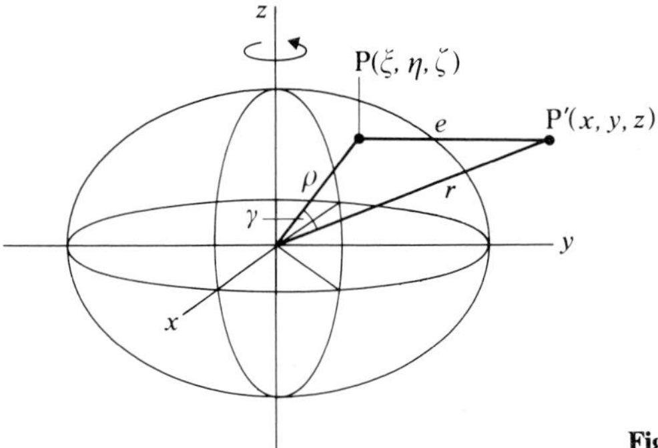

Figure 4.2 Explanation of notation.

The first term in the parentheses of Equation 4.12 is the mass of the Earth, and is

$$\int \mathrm{d}m = M.$$

As to the second term in Equation 4.12, since

$$P_1 = \cos \gamma = \frac{x\xi + y\eta + z\zeta}{\rho r},$$

$\int P_1 \rho \, \mathrm{d}m$ can be written as

$$\int P_1 \rho \, \mathrm{d}m = \frac{1}{r} \left(x \int \xi \, \mathrm{d}m + y \int \eta \, \mathrm{d}m + z \int \zeta \, \mathrm{d}m\right).$$

But since the origin of the co-ordinates is taken at the centre of the Earth,

$$\int \xi \, \mathrm{d}m = \int \eta \, \mathrm{d}m = \int \zeta \, \mathrm{d}m = 0$$

and therefore

$$\int P_1 \rho \, \mathrm{d}m = 0.$$

As to the third term in the parentheses of Equation 4.12, since

$$P_2 = \tfrac{3}{2} \cos^2 \gamma - \tfrac{1}{2},$$

$$\frac{\xi}{\rho} = \cos \varphi \cos \lambda,$$

$$\frac{\eta}{\rho} = \cos\varphi \sin\lambda,$$

$$\frac{\zeta}{\rho} = \sin\varphi,$$

and remembering also

$$\int \xi\eta \, dm = \int \eta\zeta \, dm = \int \zeta\xi \, dm = 0$$

by symmetry, $\int P_2\rho^2 \, dm$ is transformed to

$$\int P_2\rho^2 \, dm = \tfrac{3}{2}\cos^2\varphi \cos^2\lambda \int \xi^2 \, dm + \tfrac{3}{2}\cos^2\varphi \sin^2\lambda \int \eta^2 \, dm$$

$$+ \tfrac{3}{2}\sin^2\varphi \int \zeta^2 \, dm - \tfrac{1}{2}\int \xi^2 \, dm - \tfrac{1}{2}\int \eta^2 \, dm - \tfrac{1}{2}\int \zeta^2 \, dm.$$

Now

$$\int (\xi^2 + \eta^2) \, dm, \int (\eta^2 + \zeta^2) \, dm, \int (\zeta^2 + \xi^2) \, dm$$

are the moments of inertia of the Earth around the z-, x- and y-axes and, if they are denoted by C, A and A respectively,

$$\int P_2\rho^2 \, dm$$

becomes

$$\int P_2\rho^2 \, dm = \tfrac{3}{2}(A - C)(\sin^2\varphi - \tfrac{1}{3}).$$

Finally, the potential on the geoid becomes

$$W = \frac{GM}{\rho}\left\{1 + \frac{K}{2\rho^2}(1 - 3\sin^2\varphi) + \frac{\omega^2\rho^3}{2GM}\cos^2\varphi\right\} \tag{4.13}$$

where

$$K = \frac{A - C}{M}.$$

Now g is the gradient of W, which in our approximation can be written as

$$g = -\frac{\partial W}{\partial \rho}.$$

Therefore

$$g = \frac{GM}{\rho^2}\left\{1 + \frac{3K}{2\rho^2}(1 - 3\sin^2\varphi) - \frac{\omega^2\rho^3}{2GM}\cos^2\varphi\right\}. \tag{4.14}$$

The geoid is the surface of equal potential and if this potential is denoted by W_0,

$$W_0 = \frac{GM}{\rho}\left\{1 + \frac{K}{2\rho^2}(1 - 3\sin^2\varphi) + \frac{\omega^2\rho^3}{2GM}\cos^2\varphi\right\} \tag{4.15}$$

or

$$\rho = \frac{GM}{W_0}\left\{1 + \frac{K}{2\rho^2}(1 - 3\sin^2\varphi) + \frac{\omega^2\rho^3}{2GM}\cos^2\varphi\right\} \tag{4.16}$$

approximately. The gravity value at a latitude φ on the geoid is obtained if this ρ is put in the expression of g,

$$g_\varphi = \frac{W_0^2}{GM}\left\{1 + \frac{1}{2}\frac{K}{a^2} - \frac{3\omega^2a^3}{GM}\right\}\left\{1 + \left(\frac{2\omega^2a^3}{GM} - \frac{3K}{2a^2}\right)\sin^2\varphi\right\}. \tag{4.17}$$

The factor

$$\frac{W_0^2}{GM}\left\{1 + \frac{1}{2}\frac{K}{a^2} - \frac{3\omega^2a^3}{GM}\right\}$$

is simply the value of g at $\varphi = 0$, that is, the value of g_e at the Equator. The above expression for g can therefore be written as

$$g_\varphi = g_e(1 + \beta\sin^2\varphi),$$

$$\beta = \frac{2\omega^2a^3}{GM} - \frac{3K}{2a^2}.$$

Similarly

$$\rho_\varphi = \frac{GM}{W_0}\left\{1 + \frac{K}{2a^2} + \frac{\omega^2a^3}{2GM}\right\}\left\{1 - \left(\frac{3K}{2a^2} + \frac{\omega^2a^3}{2GM}\right)\sin^2\varphi\right\}.$$

But

$$\frac{GM}{W_0}\left\{1 + \frac{K}{2a^2} + \frac{\omega^2 a^3}{2GM}\right\}$$

is the value of ρ_φ at $\varphi = 0$, which is a. Therefore

$$\rho_\varphi = a(1 - \varepsilon \sin^2 \varphi), \tag{4.18}$$

$$\varepsilon = \frac{3K}{2a^2} + \frac{\omega^2 a^3}{2GM}.$$

The above ρ_φ gives an approximate figure of the geoid and this surface is called a **niveau spheroid**. The shape of this niveau spheroid is very close to a revolution ellipsoid. Its deviation from the revolution ellipsoid is the largest at $\varphi = 45°$, but is only 27 m in the case of the Earth. Adding the two relations

$$\beta = \frac{2\omega^2 a^3}{GM} - \frac{3K}{2a^2}$$

and

$$\varepsilon = \frac{3K}{2a^2} + \frac{\omega^2 a^3}{2GM},$$

we get

$$\beta + \varepsilon = \frac{5}{2}\frac{\omega^2 a^3}{GM} \approx \frac{5}{2}\frac{\omega^2 a}{g_e},$$

$$\varepsilon = \frac{5}{2}\frac{\omega^2 a}{g_e} - \beta. \tag{4.19}$$

This important relation between ε and β is known as Clairaut's theorem (Alex Claud Clairaut, 1713–65).

The dependence of g_φ on the latitude φ is therefore

$$g_\varphi = g_e(1 + \beta \sin^2 \varphi) \tag{4.20}$$

$$= g_e\left\{1 + \left(\frac{5}{2}\frac{\omega^2 a}{g_e} - \varepsilon\right) \sin^2 \varphi\right\}.$$

If the coefficient β can once be found from gravity observations, the ellipticity ε of the geoid can be calculated from

$$\varepsilon = \frac{5}{2}\frac{\omega^2 a}{g_e} - \beta.$$

As was explained before, according to Newton's calculations

$$\varepsilon = 1/231,$$

$$\frac{g_p}{g_e} = \frac{1 + \varepsilon/5}{1 - 4\varepsilon/5} \approx 1 + \varepsilon = 1.0043,$$

while according to Huygens' calculations

$$\varepsilon = 1/578,$$

$$\frac{g_p}{g_e} = \frac{1 + 2\varepsilon}{1 - 2\varepsilon} = 1 + 4\varepsilon = 1.0069.$$

Newton's g_p/g_e is smaller than Huygens' g_p/g_e and Newton's ε is larger than Huygens' ε. Clairaut's theorem says that the larger the value of β, the smaller is the value of ε because their sum $5\omega^2 a/2g_e$ is a constant. In words, the larger the rate of increase of gravity value with latitudes, the smaller is the ellipticity of the geoid and the closer it is to a sphere. This may appear at first to be a rather paradoxical statement, but considering that some adjustment of mass distribution must take place within the Earth in order to make the geoid an equipotential surface, this is not the case at all.

In Japan, the expression for g on land is

$$g_\varphi = 977.984(1 + 0.005538 \sin^2 \varphi)$$

in which the coefficient β of $\sin^2 \varphi$ is very large. This yields a very small ellipticity of the geoid 1/319.5. The geoid in the neighbourhood of Japan is much closer to a sphere than elsewhere. This is caused by an anomalous underground-mass distribution beneath this country.

If we go to a higher approximation and include the term $\sin^2 2\varphi$, g_φ and ρ_φ are given by

$$g_\varphi = g_e(1 + \beta \sin^2 \varphi - \beta' \sin^2 2\varphi), \tag{4.21}$$

$$\rho_\varphi = a(1 - \varepsilon \sin^2 \varphi - \varepsilon' \sin^2 2\varphi)$$

where

$$\beta = \frac{5}{2}\frac{\omega^2 a}{g_e} - \varepsilon - \frac{17}{14}\frac{\omega^2 a}{g_e}\varepsilon,$$

$$\beta' = \frac{\varepsilon}{8}\left(\frac{5\omega^2 a}{g_e} - \varepsilon\right),$$

$$\varepsilon' = \frac{\varepsilon}{4}\left(\frac{g_p - g_e}{g_e} - \varepsilon\right).$$

The up-to-date numerical values of the constants in this formula according to the results of observations are

$$g_\varphi = 978.03185(1 + 0.0053024 \sin^2 \varphi - 0.0000059 \sin^2 2\varphi)$$

and it gives an ellipticity

$$\varepsilon = 1/298.25.$$

In this approximation, the shape of the niveau spheroid is even closer to an ellipsoid, the largest deviation being only 17 m at $\varphi = 45°$. Somigliana (1930) of Italy showed that the formula for g can be written as follows without approximation:

$$g_\varphi = \frac{a g_e \cos^2 \varphi + b g_p \sin^2 \varphi}{\sqrt{(a^2 \cos^2 \varphi + b^2 \sin^2 \varphi)}}. \tag{4.22}$$

More recently, it has been possible to calculate the shape of the geoid by using observations of the orbits of artificial satellites. Kozai (1969) and others found that the geoid is not symmetrical with respect to the equatorial plane and that it has a shape like a pear. Near the North Pole, the geoid is 13.5 m higher and near the South Pole 24.1 m lower than an ellipsoid.

References and further reading

Clairaut, A. C. 1743. *Theorie de la figure de la terre.* Paris: Chez Courier.

Kozai, Y. 1961a. *Potential field of the Earth derived from motions of artificial satellites.* Proc. Symp. Geod. Space Age, Ohio State University, 174.

Kozai, Y. 1961b. The gravitational field of the Earth derived from motions of three satellites. *Astr. J.* **66**, 8.

Kozai, Y. 1961c. Tesseral harmonics of the gravitational potential from satellite motions. *Astr. J.* **66**, 355.

Kozai, Y. 1964. *New determination of zonal harmonics coefficients of the Earth's gravitational potential*. Spec. Rep. Smithsonian Astr. Observ., no. 165.

Kozai, Y. 1966. The Earth's gravitational potential derived from satellite motion. *Space Sci. Res.* **5**, 818.

Kozai, Y. 1969. Revised values for coefficients of zonal spherical harmonics in the geopotential. In *Dynamics of satellites*, 104. Berlin: Springer.

Somigliana, C. 1930. Sul Campo Gravitazionale Estero del Geoid Ellissoidico. *Atti Acad. Nazl, Lincei Rend.* **11**, 237.

5 Underground masses and resulting gravity anomalies

5.1 Gravity interpretation

The value of gravity measured at a point on the Earth and reduced to sea level does not usually agree with that calculated by the standard gravity formula for the latitude of that point. The difference between the two is the gravity anomaly, and it represents the downward component of the force of attraction produced by underground masses which are not visible. The process of finding out the distribution of underground masses responsible for given gravity anomalies is called gravity interpretation. There are two approaches to this: the indirect method and the direct method. In the indirect method, an underground-mass distribution which is considered probable is tentatively assumed and the resulting gravity due to this distribution calculated. Then the calculated gravity variations are compared with the observed values. If the two sets of gravity values do not agree satisfactorily, the assumed mass distribution is modified appropriately so that a better agreement is obtained. The direct method, on the other hand, aims at finding the underground mass distribution directly from the observed gravity anomaly.

This chapter deals with the indirect method only. The direct method is an application of potential theory and it will be explained in Chapter 7 after necessary mathematical preparation for it is given in Chapter 6.

In the following sections, several simple models of anomalous mass distribution are selected and the resulting gravity anomaly distribution is calculated. To simplify sentences, the word 'anomaly' or 'anomalous' will not be used each time hereafter and instead of saying 'to calculate gravity anomaly due to an anomalous mass distribution', we will simply say 'to calculate gravity distribution due to an underground mass distribution'.

5.2 Gravity due to masses of simple geometrical forms

5.2.1 *Sphere*

The attraction f of a spherical mass M is the same as that which occurs when its total mass is concentrated at its centre; f does not vary with the

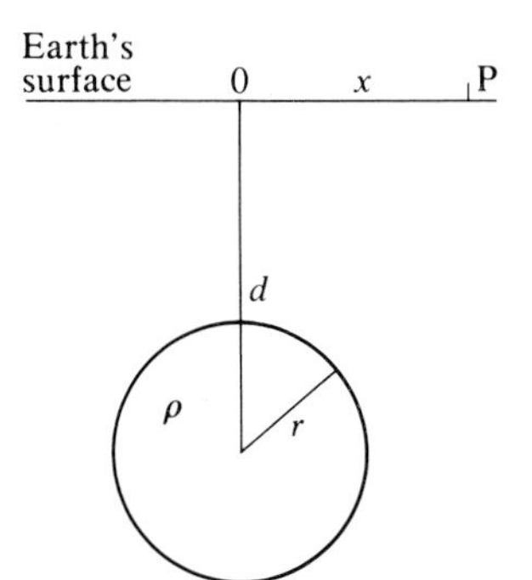

Figure 5.1 Gravity on the Earth's surface due to a spherical mass at a depth d.

density ρ nor with the radius a of the sphere, providing the mass $M = 4\pi\rho a^3/3$ is the same. In a case like this, where ρ and a cannot be determined separately, we say that the problem does not have a unique solution.

The value $g(x)$ on the Earth's surface due to a spherical mass M at a depth d (Fig. 5.1) is the downward component of its attraction $f(x)$ and is therefore

$$g(x) = f(x)\frac{d}{(d^2 + x^2)^{1/2}}$$

$$= GM\frac{1}{d^2 + x^2}\frac{d}{(d^2 + x^2)^{1/2}},$$

$$= GM\frac{d}{(d^2 + x^2)^{3/2}}. \tag{5.1}$$

At $x = 0$,

$$g(0) = GM\frac{1}{d^2}.$$

If $\rho = 1.0$, $a = 1$ km, the depth d of the sphere's centre to produce an attraction of 1 mgal at the point 0 is 5.28 km from the relation

$$10^{-3} = \tfrac{4}{3}\pi G \times \frac{10^{15}}{d^2}.$$

The ratio $g(x) : g(0)$ changes with x according to the relation

$$\frac{g(x)}{g(0)} = \frac{d^3}{(d^2 + x^2)^{3/2}} = \frac{1}{\{1 + (x/d)^2\}^{3/2}}. \tag{5.2}$$

This is a function of the ratio $x : d$ only. This means that if the horizontal distance x is expressed as a multiple of the depth of the sphere's centre, decrease of $g(x) : g(0)$ is the same whether the force of attraction is produced by a deep-seated sphere or by a shallow-seated one. The value of

$$\frac{1}{\{1 + (x/d)^2\}^{3/2}}$$

changes with $x : d$ as shown by the curve in Figure 5.2. It is 0.5 at $x/d = 0.77$, so that if the value of $g(x)$ at a point is a half of that at the origin, then the depth to the centre of the sphere is given by

$$d = \frac{x}{0.77} = 1.30\,x$$

provided, of course, that we know that the underground mass is a sphere. For large x, $g(x)$ decreases according to $1/x^3$.

5.2.2 *Horizontal line mass*

Gravity at a point P right above the middle point M of a horizontal line mass of length $2a$ at a depth d (Fig. 5.3) is

$$g_p = G\rho \int_{-a}^{a} \frac{d}{(x^2 + d^2)^{3/2}}\,dx$$

$$= \frac{2G\rho}{d} \frac{1}{\sqrt{\{1 + (d/a)^2\}}}, \qquad (5.3)$$

where ρ is the linear density of the line mass. For large a,

$$\frac{1}{\sqrt{\{1 + (d/a)^2\}}}$$

tends rapidly to 1, as shown in Figure 5.4. If a is larger than $3d$, g_p is almost equal to that value when a is infinitely large. This is very convenient in treating practical gravity problems. In the limiting case of $a = \infty$, g_p becomes $2G\rho/d$ which is inversely proportional to the depth of the line mass.

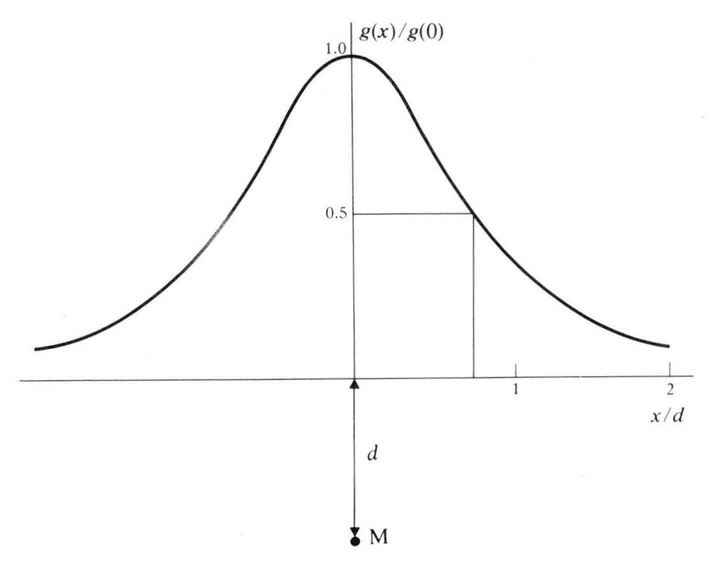

Figure 5.2 Change of $g(x)/g(0)$ according to x/d. At $x/d = 0.77$, $g(x)/g(0)$ is $\frac{1}{2}$.

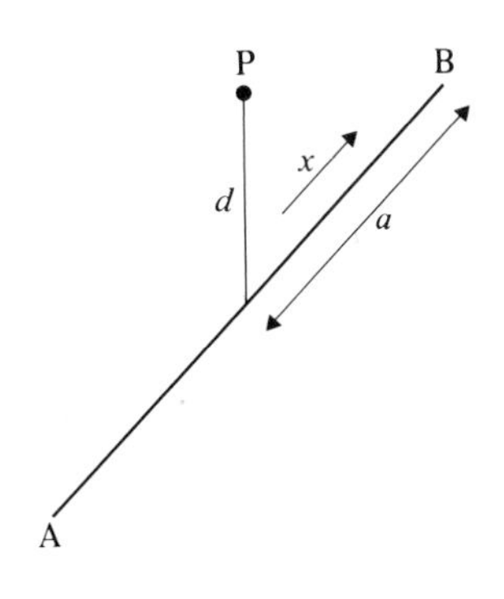

Figure 5.3 Gravity at a point P above the middle point of a horizontal line mass at a depth d.

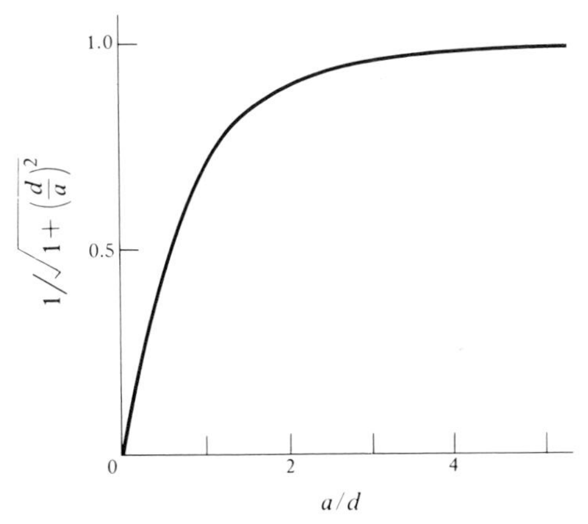

Figure 5.4 Change of $1/\sqrt{\{1 + (d/a)^2\}}$ according to a/d.

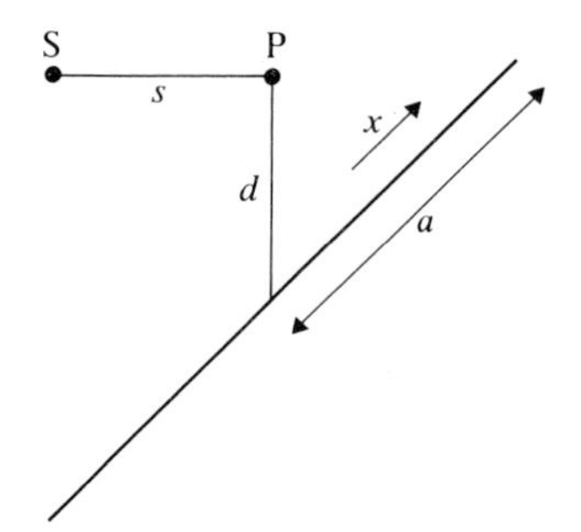

Figure 5.5 Gravity due to an infinitely long line mass at a point S, at a horizontal distance s from point P.

Gravity due to this infinitely long line mass at a point S, which is at a distance s sideways from P (Fig. 5.5), is

$$g(s) = 2G\rho \frac{d}{s^2 + d^2} \,. \tag{5.4}$$

5.2.3 *Semi-infinite horizontal plate*

Gravity at a point P due to a semi-infinite horizontal plate extending infinitely in the direction of x from $x = a$ to $x = \infty$ and also in the direction perpendicular to the plane of paper (Fig. 5.6) is

$$\begin{aligned} g_{\mathrm{p}} &= G\rho \int_a^{\infty} \frac{d}{d^2 + x^2}\,\mathrm{d}x \\ &= 2G\rho \tan^{-1}\frac{d}{a} \\ &= 2G\rho\alpha. \end{aligned} \tag{5.5}$$

α is the dip angle of the edge of the plate as seen from the point P. If this dip angle is the same, the gravity at P is also the same whether it is due to a shallower plate A or to a deeper plate C (Fig. 5.7). The curves of horizontal distribution of gravity due to plates at various depths pass the same point but at other points the curves become flatter as the depth is larger (see Fig. 5.8).

5.2.4 *Infinitely long horizontal plate with a finite width*

From what has been seen in the preceding section, gravity at a point P due to an infinitely long horizontal plate with a finite width c (Fig. 5.9) is equal to the difference

(g due to a plate extending from $x = a$ to $x = \infty$) $-$ (g due to a plate extending from $x = a + c$ to $x = \infty$) $= 2G\rho(\alpha - \beta)$.

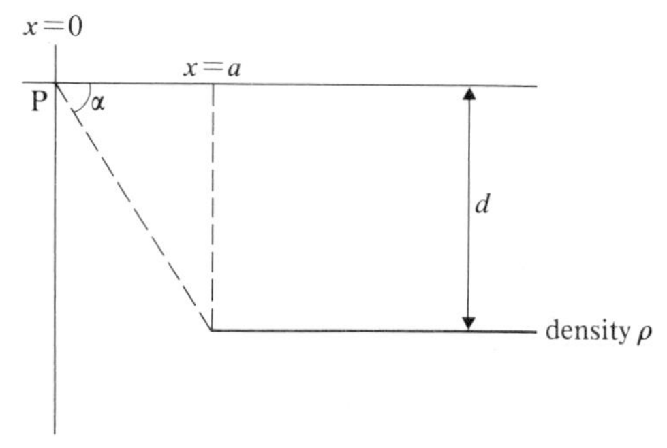

Figure 5.6 Gravity at a point P due to a semi-infinite horizontal plate extending from $x = 0$ to $x = \infty$ at a depth d.

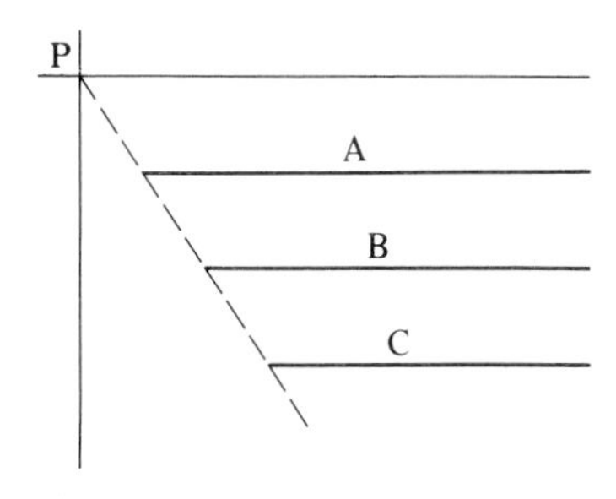

Figure 5.7 Gravity at P is the same whether it is due to a shallower plate A or a deeper plate C, if the dip angle is the same.

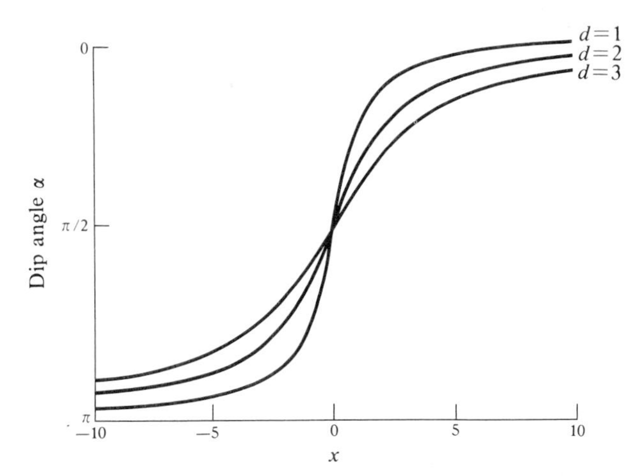

Figure 5.8 Change of dip angle according to x.

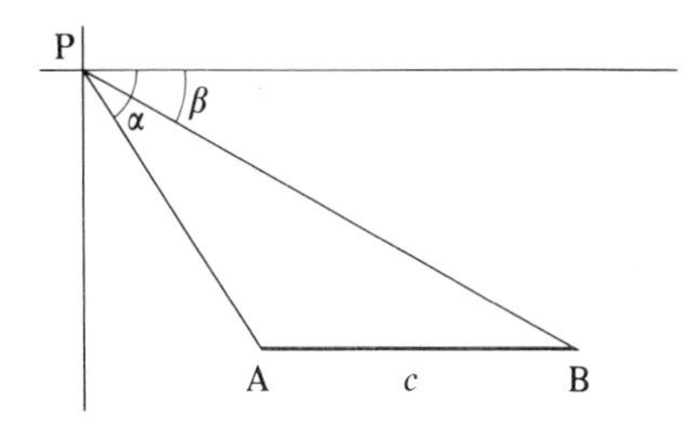

Figure 5.9 Gravity at a point P due to an infinitely long horizontal plate AB with a finite width c.

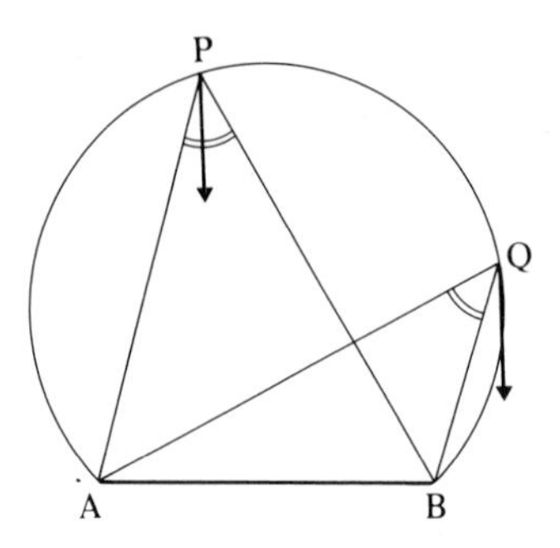

Figure 5.10 Gravity is the same everywhere on the circle.

$(\alpha - \beta)$ is the angle that is included by AB at P.

If a circle is drawn with AB as a chord, as shown in Figure 5.10, the gravity value is the same at every point on this circle, because the circumferential angle is the same on this circle. If a series of circles is drawn, on each of which AB has different circumferential angles, and if at the intersections of the circles with a horizontal line the circumferential angles are plotted, the curve passing through these plots will give the distribution of gravity along the straight line which corresponds to the Earth's surface. Above the central line of the plate, gravity is

$$g_{\mathrm{p}} = 4G\rho \tan^{-1}\frac{a}{d} = 4G\rho\alpha. \tag{5.6}$$

Figure 5.11 shows how the angle α changes according to a/d. It tends to $\pi/2$ when a/d tends to ∞. If a/d is larger than 5, the value of g is nearly 90% of that when a/d is ∞. In this limiting case

$$g = 2\pi G\rho. \tag{5.7}$$

5.2.5 *Circular plate*

Gravity at a point P right above the centre of a circular plate with a radius a (Fig. 5.12) is

$$g_{\mathrm{p}} = 2\pi G\rho \int_0^a \frac{d}{(d^2 + r^2)^{3/2}}\,\mathrm{d}r$$

$$= 2\pi G\rho\left[1 - \frac{1}{\surd\{1 + (a/d)^2\}}\right]. \tag{5.8}$$

When a tends to ∞, g will be

$$g = 2\pi G\rho.$$

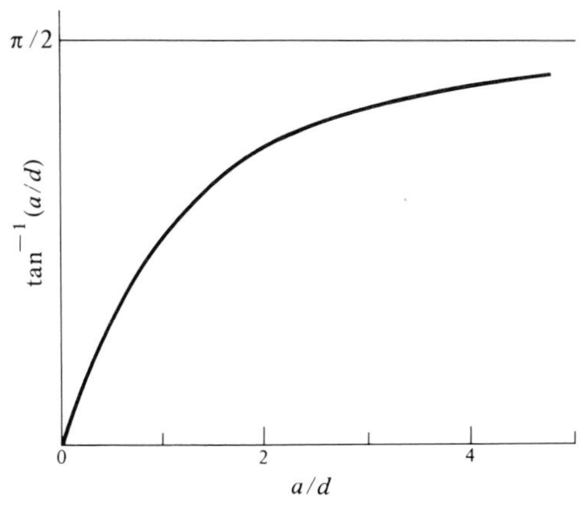

Figure 5.11 Change of $\tan^{-1}(a/d)$ according to (a/d).

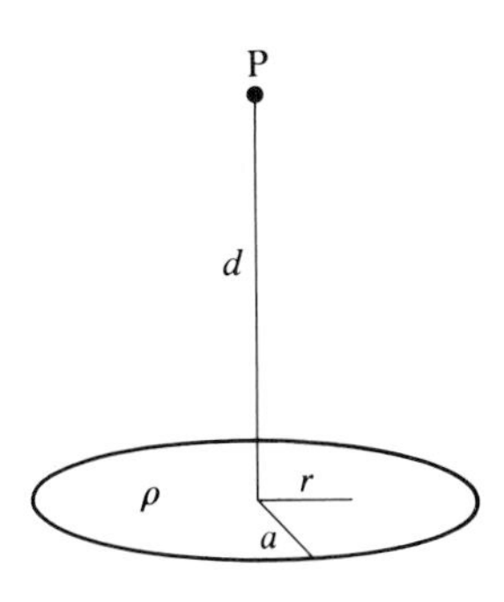

Figure 5.12 Gravity at a point P above the centre of a circular plate with a radius a at a depth d.

The factor

$$\left[1 - \frac{1}{\sqrt{\{1 + (a/d)^2\}}}\right]$$

increases rather slowly with a/d and the value of this factor at $a/d = 5$ is only 80% of the limiting value.

5.2.6 *Infinitely long horizontal bar with an arbitrary cross-sectional form*

The force of attraction due to an infinitely long horizontal bar having an arbitrary cross-sectional form can easily be calculated, by regarding this bar as a pile of several long thin plates (Fig. 5.13). Divide the cross section into a number of thin horizontal sheets and to each of them apply the relation

$$g = 2G\rho(\alpha - \beta) \tag{5.9}$$

and add the values of g for all the sheets. The sum is g_P. To facilitate this, the present author (Tsuboi 1943) made and used a convenient instrument which is shown in Figure 5.14. Put the centre of the ratchet wheel on the point P where the gravity value is to be found and place the lever L at the right end of the first sheet. Move the lever in the direction of the arrow to the left end L′ of the sheet. The wheel will be turned through the angle $(\alpha - \beta)$ by the action of the pawl R attached to the lever. Then turn the lever back to the right end of the next sheet. In this case, another pawl R′ will prevent the wheel from turning. Repeat this process until all the sheets are scanned. The total angle of wheel turn is proportional to the gravity value at P.

5.2.7 *Two parallel straight lines*

Gravity due to two parallel straight line masses at a point P above the middle of the lines (Fig. 5.15) is

$$g_P = 2G\rho\left\{\frac{d}{(a+x)^2+d^2} + \frac{d}{(a-x)^2+d^2}\right\}. \tag{5.10}$$

The upper curve in Figure 5.16 corresponds to the case of $d = a$ and the lower curve to the case of $d = 2a$. In the former case, two maxima are seen in gravity distribution while in the latter case, only one maximum is seen. This is a good example showing that one maximum of gravity does not necessarily imply one maximum in the underground mass distribution. In this case of two parallel line masses, the depth of the lines at which they cease to show two maxima of gravity is that at which the curvature of the curve of gravity at $x = 0$ is zero. From

$$\left.\frac{d^2g}{dx^2}\right|_{x=0} = 0 = 2G\rho\frac{-4d(1+d^2)+16d}{(1+d)^3} \tag{5.11}$$

we obtain

$$d = a\sqrt{3}.$$

5.3 Optical analogy

Gravity at a point P due to a plane mass M is proportional to

$$\int \frac{\cos\theta}{r^2}\, dS$$

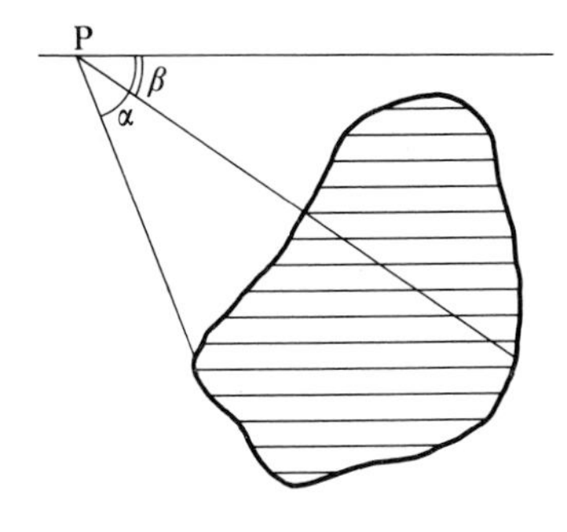

Figure 5.13 Gravity due to an infinitely long horizontal bar of arbitrary cross-sectional form.

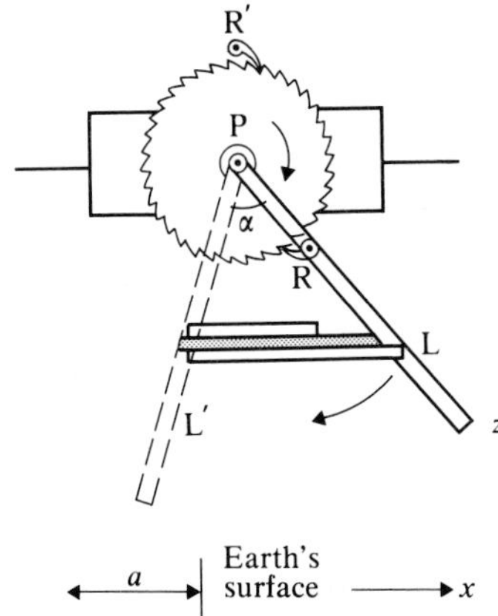

Figure 5.14 An instrument for measuring the sum of α.

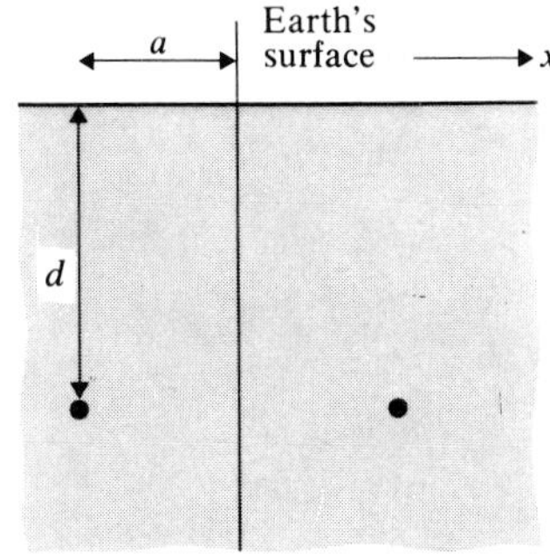

Figure 5.15 Gravity due to two line masses at a depth d with a separation $2a$.

as shown in Figure 5.17. If M is regarded as a plane light source of uniform luminosity, the intensity of illumination on the plane OP is also proportional to

$$\int \frac{\cos\theta}{r^2}\,\mathrm{d}S.$$

Using this optical analogy, it is possible to get an approximate idea of what the gravity distribution is like due to a plane mass of arbitrary shape (Tsuboi 1938). Cut out the shape from a sheet of black paper and place the remaining sheet on a milk-white glass plate. Place a light-

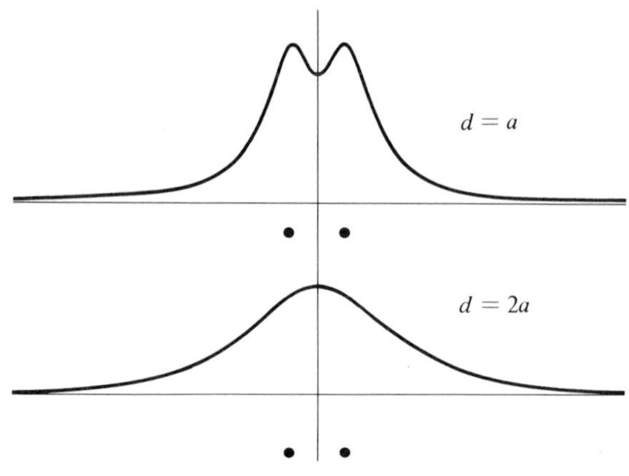

Figure 5.16 Gravity due to the two line masses at $d = a$ and $d = 2a$. The curve for $d = a$ has two maxima, but that for $d = 2a$ has only one maximum.

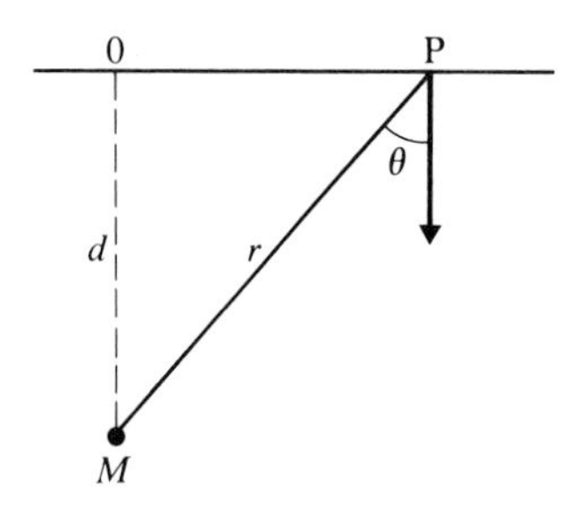

Figure 5.17 Gravity at point P is proportional to $\int \cos\theta/r^2\, ds$.

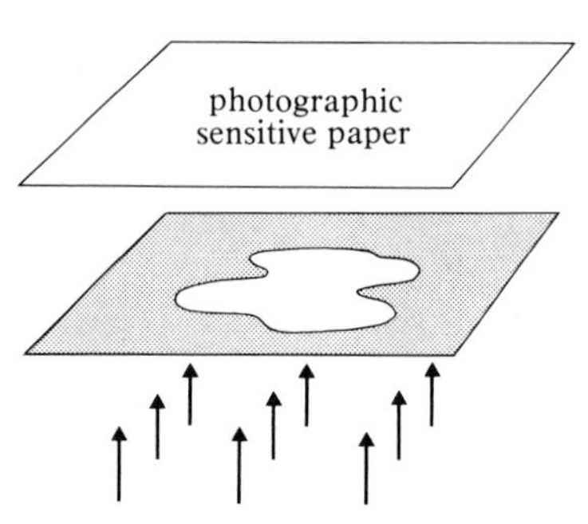

Figure 5.18 Apparatus used to obtain approximate idea of gravity distribution. Light is projected from below to illuminate the plate; the distribution of blackening on the photographic paper is similar to that of gravity on the Earth's surface.

sensitive paper at a certain height from the glass plate and illuminate the plate uniformly from below. The distribution of blackening on the photographic paper is similar to that of gravity on the Earth's surface produced by the mass having the shape which is cut off. Figure 5.19 is an example in which an L-shape is cut out and the photographic paper is placed at various heights from the glass plate. It is clearly seen that the nearer the photographic paper to the plate, the sharper is the boundary of the L-shape; and the more distant the paper is from the glass plate,

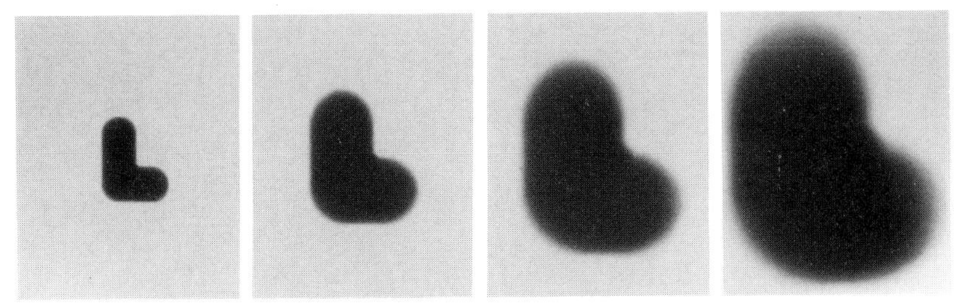

Figure 5.19 Blackening of photographic papers at various heights.

the more obscure are the boundary outlines. The contrast of the black part with the background also changes according to the distance of the photographic paper and the glass plate. This pattern is the same as that which the gravity values will show due to an L-shaped underground mass at various depths.

References and further reading

Talwani, M. and M. Ewing 1960. Rapid computation of gravitational attraction of three-dimensional bodies of arbitrary shape. *Geophys.* **25**, 203.

Tsuboi, C. 1943. Calculation of gravity due to an infinitely long column with an arbitrary cross-sectional form (in Japanese). *J. Seis. Soc., Japan* **15**, 167.

6 Potential theory of gravity

6.1 Gravity potential

In this chapter, the concept of potential will be introduced as a preparation for advanced discussions of the relationships between an underground-mass distribution and the resulting gravity distribution. By using the concept of potential, a number of problems related to gravity can be dealt with in a more general and systematic way than by referring to specific model calculations as we have done in Chapter 5. The potential specifically related to gravity force is called the **gravity potential**.

When a mass is moved from a point A to another point B in a space in which no force is acting on the mass other than the attraction due to a certain body M, the amount of work V needed for this movement does not depend on its path between the points A and B and is the same whichever path is followed. More especially, if the point B is considered to be at an infinite distance, the value of V is a function of the original position of the point A only. We will call this V the potential at A. The potential as defined here is the negative of the potential energy, which is the amount of work needed for bringing a mass from infinity to that point and which is usually used in dynamics.

If there is a point mass M at $r = 0$, the work needed to move a unit mass from a point at a distance r from M to infinity against the attraction due to M is

$$V = \int_r^\infty \frac{GM}{r^2}\,\mathrm{d}r = \left| -\frac{GM}{r} \right|_r^\infty = \frac{GM}{r} \qquad (6.1)$$

according to our definition and is a function of r only. This is a positive quantity. According to this definition of potential, the force f working in the direction of increasing r is given simply by differentiating V with respect to r. In our case, the force acting in the direction of increasing r is

$$\frac{\partial V}{\partial r} = -\frac{GM}{r^2}\,. \qquad (6.2)$$

The minus sign in front of GM/r^2 indicates that the force is working in the direction of decreasing r, which is towards M.

Generally, if V is differentiated in any direction, that will at once give the force acting in that direction. The force acts in the direction in which the potential increases. It is because of this simplicity and convenience that the potential V as defined here will be used in this book.

6.2 Laplace's equation

Let V be the potential at a point. If there is no mass at this point, the following relation holds:

$$\frac{\partial^2 V}{\partial x^2} + \frac{\partial^2 V}{\partial y^2} + \frac{\partial^2 V}{\partial z^2} = 0. \qquad (6.3)$$

This is called Laplace's equation (Pierre Simon Laplace, 1749–1827). The operator

$$\frac{\partial^2}{\partial x^2} + \frac{\partial^2}{\partial y^2} + \frac{\partial^2}{\partial z^2}$$

is called a Laplacian and is often denoted by ∇^2 or Δ. Using this notation, Laplace's equation may be written as

$$\nabla^2 V = 0 \quad \text{or} \quad \Delta V = 0.$$

This equation is the fundamental basis for discussing many gravity problems. It can be derived as follows. The space in Figure 6.1 is a field of attraction force due to a certain mass body not shown in the figure. S is a closed surface drawn in this space. The curved lines in the figure are the lines of attraction force, drawn in the direction of the attraction force f working at each point. If no mass is enclosed within the surface S, all the lines of force that come into the space enclosed by the surface go out of it. In other words, incoming and outgoing lines of force are equal in number and therefore their total sum must be zero with due

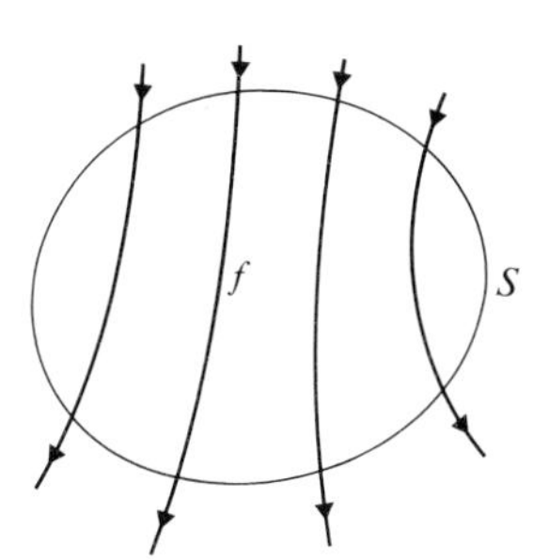

Figure 6.1 A field of attraction force. If no mass is enclosed within the surface S, all the lines of force f which come in across S go out of it.

regard to algebraic signs. If **n** is a unit vector drawn in the direction of the outward normal to the surface S, then the scalar produce $\mathbf{f}\cdot\mathbf{n}$ {(the product of the magnitudes of the vectors **f** and **n**) × (cosine of the angle between **f** and **n**)} integrated over the whole surface of S must be zero, that is

$$\iint \mathbf{f}\cdot\mathbf{n}\,\mathrm{d}S = 0. \tag{6.4}$$

In vector analysis, there is a theorem by which the surface integral given above can be transformed into a volume integral as follows:

$$\iint \mathbf{f}\cdot\mathbf{n}\,\mathrm{d}S = \iiint \left(\frac{\partial \mathbf{f}_x}{\partial x} + \frac{\partial \mathbf{f}_y}{\partial y} + \frac{\partial \mathbf{f}_z}{\partial z}\right) \mathrm{d}v,$$

where $\mathbf{f}_x$, $\mathbf{f}_y$ and $\mathbf{f}_z$ are the x-, y- and z-components of the force **f** respectively.

$$\left(\frac{\partial \mathbf{f}_x}{\partial x} + \frac{\partial \mathbf{f}_y}{\partial y} + \frac{\partial \mathbf{f}_z}{\partial z}\right)$$

in the above expression is called the divergence of the vector **f** and is written as

$$\frac{\partial \mathbf{f}_x}{\partial x} + \frac{\partial \mathbf{f}_y}{\partial y} + \frac{\partial \mathbf{f}_z}{\partial z} = \operatorname{div} \mathbf{f}. \tag{6.5}$$

Using this notation

$$\iint \mathbf{f}\cdot\mathbf{n}\,\mathrm{d}S = \iiint \operatorname{div} \mathbf{f}\,\mathrm{d}v = 0.$$

If the force **f** in question is derived from a potential V so that

$$\mathbf{f}_x = \frac{\partial V}{\partial x} \qquad \mathbf{f}_y = \frac{\partial V}{\partial y} \qquad \mathbf{f}_z = \frac{\partial V}{\partial z},$$

then the above expression becomes

$$\iiint \left(\frac{\partial^2 V}{\partial x^2} + \frac{\partial^2 V}{\partial y^2} + \frac{\partial^2 V}{\partial z^2}\right) \mathrm{d}v = 0.$$

But since the volume element dv can be taken arbitrarily and there is no condition imposed about how it should be chosen, the integrand in the above equation itself must be zero everywhere and

$$\frac{\partial^2 V}{\partial x^2} + \frac{\partial^2 V}{\partial y^2} + \frac{\partial^2 V}{\partial z^2} = 0. \tag{6.6}$$

This is simply Laplace's equation.

6.3 Meaning of Laplace's equation

The meaning of Laplace's equation will perhaps be better understood by the following considerations. Let us take a small sphere of radius a with a point O at its centre. Expanding the potential V in the neighbourhood of O into a Taylor's series, we obtain

$$\begin{aligned} V(xyz) = V_0 &+ x\frac{\partial V}{\partial x} + y\frac{\partial V}{\partial y} + z\frac{\partial V}{\partial z} \\ &+ \tfrac{1}{2}x^2\frac{\partial^2 V}{\partial x^2} + \tfrac{1}{2}y^2\frac{\partial^2 V}{\partial y^2} + \tfrac{1}{2}z^2\frac{\partial^2 V}{\partial z^2} \\ &+ xy\frac{\partial^2 V}{\partial x\,\partial y} + yz\frac{\partial^2 V}{\partial y\,\partial z} + zx\frac{\partial^2 V}{\partial z\,\partial x} \\ &+ \cdots . \end{aligned} \tag{6.7}$$

The mean value of V within this small sphere is

$$\bar{V} = \frac{1}{4\pi a^3/3}\iiint V\,\mathrm{d}v.$$

From the Taylor series expansion form of $V(x\ y\ z)$, we see that

$$\iiint V\,\mathrm{d}v = \underset{(1)}{\iiint V_0\,\mathrm{d}v} + \underset{(2)}{\frac{\partial V}{\partial x}\iiint x\,\mathrm{d}v} + \underset{(3)}{\frac{\partial V}{\partial y}\iiint y\,\mathrm{d}v} + \underset{(4)}{\frac{\partial V}{\partial z}\iiint z\,\mathrm{d}v}$$

$$+ \underset{(5)}{\frac{1}{2}\frac{\partial^2 V}{\partial x^2}\iiint x^2\,\mathrm{d}v} + \underset{(6)}{\frac{1}{2}\frac{\partial^2 V}{\partial y^2}\iiint y^2\,\mathrm{d}v} + \underset{(7)}{\frac{1}{2}\frac{\partial^2 V}{\partial z^2}\iiint z^2\,\mathrm{d}v}$$

$$+ \underset{(8)}{\frac{\partial^2 V}{\partial x \partial y} \iiint xy \,\mathrm{d}v} + \underset{(9)}{\frac{\partial^2 V}{\partial y \partial z} \iiint yz \,\mathrm{d}v} + \underset{(10)}{\frac{\partial^2 V}{\partial z \partial x} \iiint zx \,\mathrm{d}v}$$

$$+ \cdots . \tag{6.8}$$

The first term (1) of Equation 6.8 is

$$\iiint V_0 \,\mathrm{d}v = \tfrac{4}{3}\pi a^3 V_0.$$

The second, third, fourth, eighth, ninth and tenth terms are all zero, because of symmetry relations. As for the fifth, sixth and seventh terms, we see that

$$\iiint x^2 \,\mathrm{d}v = \iiint y^2 \,\mathrm{d}v = \iiint z^2 \,\mathrm{d}v = \tfrac{4}{15}\pi a^5.$$

The expression for $\bar{V}$ can therefore be written as

$$\bar{V} = V_0 + \frac{2\pi a^5/15}{4\pi a^3/3}\left(\frac{\partial^2 V}{\partial x^2} + \frac{\partial^2 V}{\partial y^2} + \frac{\partial^2 V}{\partial z^2}\right). \tag{6.9}$$

But Laplace's equation says that

$$\frac{\partial^2 V}{\partial x^2} + \frac{\partial^2 V}{\partial y^2} + \frac{\partial^2 V}{\partial z^2} = 0.$$

Therefore the above relation is reduced to

$$\bar{V} = V_0.$$

In words, the value of the potential V_0 at any point is equal to the mean value of V within a small sphere taken around that point. There can be no spatial maximum or minimum in the space. This is what Laplace's equation means. If, in the neighbourhood of the Earth's surface, V is the gravity potential and if Laplace's equation for V is differentiated with respect to the downward direction z, we obtain

$$\frac{\partial}{\partial z}\left(\frac{\partial^2 V}{\partial x^2} + \frac{\partial^2 V}{\partial y^2} + \frac{\partial^2 V}{\partial z^2}\right) = 0$$

$$= \frac{\partial^2}{\partial x^2}\frac{\partial V}{\partial z} + \frac{\partial^2}{\partial y^2}\frac{\partial V}{\partial z} + \frac{\partial^2}{\partial z^2}\frac{\partial V}{\partial z}$$

$$= \frac{\partial^2 g}{\partial x^2} + \frac{\partial^2 g}{\partial y^2} + \frac{\partial^2 g}{\partial z^2} \tag{6.10}$$

because $g = \partial V/\partial z$. The gravity value also satisfies Laplace's equation and there can be no spatial maximum or minimum in its distribution, as was seen above. This may perhaps appear to be hard to understand at first, because we know that there are maxima or minima in actual gravity distributions. The following simple example will be useful in making this perplexing circumstance comprehensible.

Suppose, for instance, that there is an infinitely long horizontal straight line mass at a depth d from the ground. For simplicity, the linear density of the line mass will be taken to be 1. The gravity value at $x = x$, $z = z$ due to this line mass is

$$g_{(x\,z)} = G\,\frac{d - z}{\{(d - z)^2 + x^2\}}\,. \tag{6.11}$$

Along the Earth's surface, $z = 0$, g is $d/(d^2 + x^2)$ and is at its maximum at $x = 0$. We see that, at $x = 0$, $z = 0$,

$$\frac{\partial g}{\partial x} = G\,\frac{-2x(d - z)}{\{(d - z)^2 + x^2\}^2} = 0$$

but

$$\frac{\partial g}{\partial z} = G\,\frac{(d - z)^2 - x^2}{\{(d - z)^2 + x^2\}^2} = G\,\frac{1}{d^2}$$

is not zero, so that g at $x = 0$ and $z = 0$ is not a spatial maximum. Also at $x = 0$ and $z = 0$,

$$\frac{\partial^2 g}{\partial x^2} = G\,\frac{-2(d - z)^3 + 6x^2(d - z)}{\{(d - z)^2 + x^2\}^3} = -G\,\frac{2}{d^3}\,,$$

$$\frac{\partial^2 g}{\partial z^2} = G\,\frac{2(d - z)^3 - 6x^2(d - z)}{\{(d - z)^2 + x^2\}^3} = G\,\frac{2}{d^3}$$

and $\partial^2 g/\partial x^2$ and $\partial^2 g/\partial z^2$ are equal in magnitude but opposite in algebraic sign, so that they cancel each other and satisfy Laplace's equation. g can be a maximum at $x = 0$ only in the direction of x, but this is not a spatial maximum. The mean value of g within a small sphere is always equal to the value of g at its centre.

6.4 Gauss's theorem

When Laplace's equation was derived in a preceding section, it was assumed that no mass was enclosed by the surface S. The equation should be modified in the case where a mass is enclosed within the surface.

Consider a closed surface S that encloses a mass m_1. In that case, not all the lines of force which come into it go out and some of them terminate at the mass. The number of such lines of force is $4\pi G$ times the mass m_1. If $\mathbf{n}$ is a unit vector drawn in the direction of an outward normal to the surface S, then the scalar product $\mathbf{f}\cdot\mathbf{n}$ integrated over the entire surface will not be zero but is

$$\iint (\mathbf{f}\cdot\mathbf{n})\,\mathrm{d}S = -4\pi G m_1. \tag{6.12}$$

A minus sign is added on the right-hand side of this equation because the unit vector is the outward normal drawn to the surface S. If, instead of $\mathbf{n}$, we use the inward normal $\mathbf{n}'$ and write $\mathbf{g}$ instead of $\mathbf{f}$, this equation becomes

$$\iint (\mathbf{g}\cdot\mathbf{n}')\,\mathrm{d}S = 4\pi G m_1.$$

If there are a number of masses, $m_1, m_2 \ldots m_n$ within the surface, then

$$\iint (\mathbf{g}\cdot\mathbf{n}')\,\mathrm{d}S = 4\pi G(m_1 + m_2 + \cdots + m_n).$$

This is called Gauss's theorem (C. F. Gauss, 1774–1855). Gauss's theorem is useful for finding the amount of mass enclosed by a surface from the gravity on the surface, even if the shape and size of the mass are not known. For example, applying the theorem to the whole Earth and considering a closed surface that encloses the Earth completely within it, the mass M of the Earth can be found as follows. If a spherical surface with radius R is taken, which closely encloses the whole Earth, and if the gravity on this spherical surface is uniformly g then, from Gauss's theorem

$$\iint g\,\mathrm{d}S = 4\pi G M.$$

But since

$$\iint g\,\mathrm{d}S = 4\pi R^2 g$$

it follows that

$$GM = R^2 g \tag{6.13}$$

or

$$M = \frac{gR^2}{G}.$$

Gauss's theorem can be used also for estimating the mass of underground natural resources. We integrate the gravity-anomaly values on the Earth's surface due to a suspected mass and divide the integral by $2\pi G$ instead of $4\pi G$. The result will give the total mass, even if its shape and size are unknown. The reason for using $2\pi G$ in this case instead of $4\pi G$ is that the integration over the whole of the enclosing surface is not possible but is done only on its half.

6.5 Laplace's equation in cylindrical and spherical co-ordinates

Laplace's equation in cartesian co-ordinates (x, y, z) is

$$\frac{\partial^2 V}{\partial x^2} + \frac{\partial^2 V}{\partial y^2} + \frac{\partial^2 V}{\partial z^2} = 0$$

The expression of the equation in co-ordinates (α, β, γ) other than cartesian can be obtained as follows. If

$$\alpha = f_1(x, y, z) \qquad x = F_1(\alpha, \beta, \gamma),$$

$$\beta = f_2(x, y, z) \qquad y = F_2(\alpha, \beta, \gamma),$$

$$\gamma = f_3(x, y, z) \qquad z = F_3(\alpha, \beta, \gamma),$$

the line element $\mathrm{d}s$ is expressed by

$$(\mathrm{d}s)^2 = (\mathrm{d}x)^2 + (\mathrm{d}y)^2 + (\mathrm{d}z)^2$$

$$= (h_1\, \mathrm{d}\alpha_1)^2 + (h_2\, \mathrm{d}\alpha_2)^2 + (h_3\, \mathrm{d}\alpha_3)^2,$$

where

$$h_1^2 = \left(\frac{\partial x}{\partial \alpha}\right)^2 + \left(\frac{\partial y}{\partial \alpha}\right)^2 + \left(\frac{\partial z}{\partial \alpha}\right)^2,$$

$$h_2^2 = \left(\frac{\partial x}{\partial \beta}\right)^2 + \left(\frac{\partial y}{\partial \beta}\right)^2 + \left(\frac{\partial z}{\partial \beta}\right)^2,$$

$$h_3^2 = \left(\frac{\partial x}{\partial \gamma}\right)^2 + \left(\frac{\partial y}{\partial \gamma}\right)^2 + \left(\frac{\partial z}{\partial \gamma}\right)^2.$$

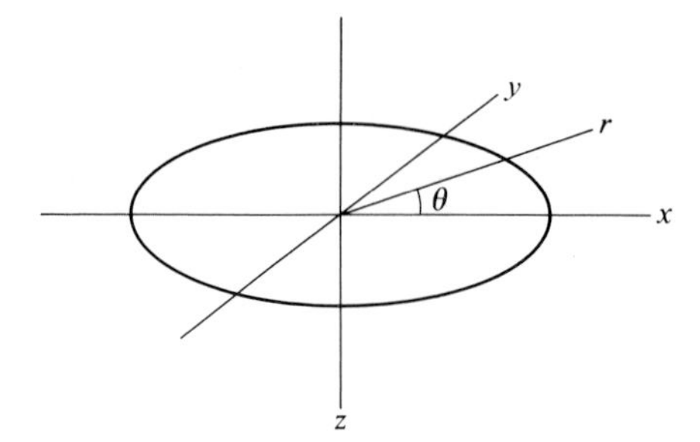

Figure 6.2 Relationship between the cartesian co-ordinates (x, y, z) and the cylindrical co-ordinates (r, θ, z).

Using these h_1, h_2, h_3, Laplace's equation in (α, β, γ) co-ordinates is

$$\frac{\partial}{\partial \alpha}\left(\frac{h_2 h_3}{h_1}\frac{\partial V}{\partial \alpha}\right)+\frac{\partial}{\partial \beta}\left(\frac{h_3 h_1}{h_2}\frac{\partial V}{\partial \beta}\right)+\frac{\partial}{\partial \gamma}\left(\frac{h_1 h_2}{h_3}\frac{\partial V}{\partial \gamma}\right)=0. \qquad (6.15)$$

This relation can be obtained by geometrical transformations of the co-ordinates.

Let us apply this transformation to cylindrical co-ordinates (r, θ, z). In this case (Fig. 6.2), the transformation relationships between (x, y, z) and (r, θ, z) are

$$r = \sqrt{(x^2 + y^2)} \qquad x = r\cos\theta,$$

$$\theta = \tan^{-1}\left(\frac{y}{x}\right) \qquad y = r\sin\theta,$$

$$z = z \qquad z = z.$$

Since the line element ds is given by

$$(ds)^2 = (dx)^2 + (dy)^2 + (dz)^2$$

$$= (dr)^2 + (r\, d\theta)^2 + (dz)^2$$

and

$$h_1{}^2 = \left(\frac{\partial x}{\partial r}\right)^2 + \left(\frac{\partial y}{\partial r}\right)^2 + \left(\frac{\partial z}{\partial r}\right)^2$$

$$= \cos^2\theta + \sin^2\theta$$

$$= 1,$$

$$h_2^2 = \left(\frac{\partial x}{\partial \theta}\right)^2 + \left(\frac{\partial y}{\partial \theta}\right)^2 + \left(\frac{\partial z}{\partial \theta}\right)^2$$

$$= r^2 \sin^2 \theta + r^2 \cos^2 \theta$$

$$= r^2$$

$$h_3^2 = \left(\frac{\partial x}{\partial z}\right)^2 + \left(\frac{\partial y}{\partial z}\right)^2 + \left(\frac{\partial z}{\partial z}\right)^2$$

$$= 1,$$

Laplace's equation in cylindrical co-ordinates is then

$$\frac{\partial}{\partial r}\left(r \frac{\partial V}{\partial r}\right) + \frac{\partial}{\partial \theta}\left(\frac{1}{r} \frac{\partial V}{\partial \theta}\right) + \frac{\partial}{\partial z}\left(r \frac{\partial V}{\partial z}\right) = 0 \qquad (6.16)$$

or

$$\frac{\partial^2 V}{\partial r^2} + \frac{1}{r}\frac{\partial V}{\partial r} + \frac{1}{r^2}\frac{\partial^2 V}{\partial \theta^2} + \frac{\partial^2 V}{\partial z^2} = 0.$$

This expression of Laplace's equation is especially useful for studying the gravity field around one particular point at $r = 0$.

For spherical co-ordinates (r, θ, ϕ) (see Fig. 6.3),

$$r = \sqrt{(x^2 + y^2 + z^2)}, \qquad x = r \sin \theta \cos \phi,$$

$$\theta = \tan^{-1} \frac{\sqrt{(x^2 + y^2)}}{z}, \qquad y = r \sin \theta \sin \phi,$$

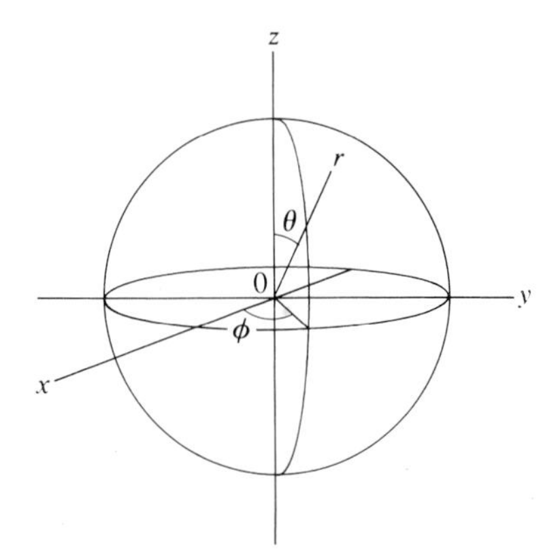

Figure 6.3 Relationship between the cartesian co-ordinates (x, y, z) and spherical co-ordinates (r, θ, ϕ).

$$\phi = \tan^{-1}\frac{y}{z}, \qquad z = r\cos\theta,$$

and

$$
\begin{aligned}
(\mathrm{d}s)^2 &= (\mathrm{d}x)^2 + (\mathrm{d}y)^2 + (\mathrm{d}z)^2 \\
&= (\mathrm{d}r)^2 + (r\,\mathrm{d}\theta)^2 + (r\sin\theta\,\mathrm{d}\phi)^2, \qquad (6.17)
\end{aligned}
$$

$$
\begin{aligned}
h_1{}^2 &= \left(\frac{\partial x}{\partial r}\right)^2 + \left(\frac{\partial y}{\partial r}\right)^2 + \left(\frac{\partial z}{\partial r}\right)^2 \\
&= \sin^2\theta\cos^2\phi + \sin^2\theta\sin^2\phi + \cos^2\theta \\
&= 1, \qquad (6.18)
\end{aligned}
$$

$$
\begin{aligned}
h_2{}^2 &= \left(\frac{\partial x}{\partial \theta}\right)^2 + \left(\frac{\partial y}{\partial \theta}\right)^2 + \left(\frac{\partial z}{\partial \theta}\right)^2 \\
&= r^2\cos^2\theta\cos^2\phi + r^2\cos^2\theta\sin^2\phi + r^2\sin^2\theta \\
&= r^2, \qquad (6.19)
\end{aligned}
$$

$$
\begin{aligned}
h_3{}^2 &= \left(\frac{\partial x}{\partial \phi}\right)^2 + \left(\frac{\partial y}{\partial \phi}\right)^2 + \left(\frac{\partial z}{\partial \phi}\right)^2 \\
&= r^2\sin^2\theta\sin^2\phi + r^2\sin^2\theta\cos^2\phi \\
&= r^2\sin^2\theta, \qquad (6.20)
\end{aligned}
$$

Laplace's equation in spherical co-ordinates is therefore

$$\frac{\partial}{\partial r}\left(r^2\sin\theta\frac{\partial V}{\partial r}\right) + \frac{\partial}{\partial \theta}\left(\sin\theta\frac{\partial V}{\partial \theta}\right) + \frac{\partial}{\partial \phi}\left(\frac{1}{\sin\theta}\frac{\partial V}{\partial \phi}\right) = 0 \qquad (6.21)$$

or

$$\frac{\partial}{\partial r}\left(r^2\frac{\partial V}{\partial r}\right) + \frac{1}{\sin\theta}\frac{\partial}{\partial \theta}\left(\sin\theta\frac{\partial V}{\partial \theta}\right) + \frac{1}{\sin^2\theta}\frac{\partial^2 V}{\partial \phi^2} = 0.$$

This equation is important for studying the gravity field over the surface of a spherical Earth. The actual Earth has an approximately rotational shape around the θ-axis, so that the potential V may be regarded to be independent of ϕ. Laplace's equation in that case simply becomes

$$\frac{\partial}{\partial r}\left(r^2 \frac{\partial V}{\partial r}\right) + \frac{1}{\sin^2 \theta}\frac{\partial}{\partial \theta}\left(\sin \theta \frac{\partial V}{\partial \theta}\right) = 0,$$

omitting the ϕ term from the above expression. In this case, it is convenient to use a new variable $\mu = \cos \theta$ instead of θ itself. Since

$$\frac{\partial}{\partial \theta} = \frac{\partial \mu}{\partial \theta}\frac{\partial}{\partial \mu} = -\sqrt{(1 - \mu^2)}\,\frac{\partial}{\partial \mu},$$

Laplace's equation will become

$$\frac{\partial}{\partial r}\left(r^2 \frac{\partial V}{\partial r}\right) + (1 - \mu^2)\frac{\partial^2 V}{\partial \mu^2} - 2\mu \frac{\partial V}{\partial \mu} = 0. \qquad (6.22)$$

Now that we have obtained Laplace's equation expressed in various co-ordinates, let us solve these equations and use the solutions for discussing the relationships existing between gravity values and the responsible underground mass distribution.

References and further reading

Byerly, W. E. 1893. *An elementary treatise on Fourier's series and spherical, cylindrical and ellipsoidal harmonics*. Boston: Ginn.
Kellog, O. D. 1929. *Foundation of potential theory*. New York: Ungar.
MacMillan, W. D. 1930. *The theory of the potential*. New York: Dover.

7 Gravity interpretation in cartesian co-ordinates

7.1 Solution of Laplace's equation in cartesian co-ordinates

In this chapter, Laplace's equation in cartesian co-ordinates will be solved and the solution will be used for finding the underground mass distribution that is responsible for a gravity distribution observed on the Earth's surface. This direct method was first developed by the present author in 1938 and is widely used now.

First of all, one-dimensional problems will be dealt with, in which both the underground mass and the resulting gravity value vary in one horizontal direction (x) only. Laplace's equation in this case is

$$\frac{\partial^2 V}{\partial x^2} + \frac{\partial^2 V}{\partial z^2} = 0,$$

z being taken vertically downwards.

To solve this equation, the potential $V(xz)$ will be assumed to be a product of $X(x)$ which is a function of x alone and $Z(z)$ which is a function of z alone:

$$V(xz) = X(x)Z(z).$$

Putting this $V(xz)$ in Laplace's equation, we obtain

$$Z\frac{\mathrm{d}^2 X}{\mathrm{d}x^2} + X\frac{\mathrm{d}^2 Z}{\mathrm{d}z^2} = 0$$

or

$$\frac{1}{X}\frac{\mathrm{d}^2 X}{\mathrm{d}x^2} + \frac{1}{Z}\frac{\mathrm{d}^2 Z}{\mathrm{d}z^2} = 0.$$

The first term on the left-hand side of the last equation is apparently a function of x alone and does not include z, while the second term is apparently a function of z alone and does not include x, but nevertheless their sum must be zero. This is possible only when the two terms are both constants, equal in magnitude but opposite in algebraic sign, so that they cancel each other. With the present sign convention, these constants must be such that

$$\frac{1}{X}\frac{d^2X}{dx^2} = -m^2$$

$$\frac{1}{Z}\frac{d^2Z}{dz^2} = m^2,$$

where m is a positive integer. The solutions to the above two equations are

$$X(x) = \frac{\cos}{\sin} mx$$

and

$$Z(z) = e^{\pm mz}$$

respectively, so that their product

$$V(xz) = X(x)Z(z) = \frac{\cos}{\sin} mx\, e^{\pm mz}$$

is a solution of Laplace's equation. Since m can be any positive integer, the general solution for $V(xz)$ is obtained by multiplying the above solution $V(xz)$ by a constant A_m and by adding up the products for all values of m:

$$V(xz) = \sum_m A_m \frac{\cos}{\sin} mx\, e^{\pm mz}. \tag{7.1}$$

There is a double algebraic sign in the exponential term of mz. The choice of the sign depends on the way of taking the positive direction of the z-axis and its origin. Here, the positive direction of the z-axis is taken downwards and the origin $z = 0$ at a depth d of a horizontal underground plane on which the mass is assumed to be distributed with varying surface densities. Then the potential V in the space $z < 0$ above the mass plane must be such that when it is differentiated with respect to z, the resulting force is positive and directed downwards. Thus we see that, in this space, $z < 0$, e^{+mz} should be taken and

$$V(xz) = \sum_m A_m \frac{\cos}{\sin} mx\, e^{+mz} \tag{7.2}$$

is an appropriate solution. In the space $z > 0$, deeper than the mass plane, the force due to it is negative and directed upwards, so that e^{-mz} should be taken and

$$V(xz) = \sum_m A_m \begin{matrix}\cos\\ \sin\end{matrix} mx e^{-mz} \tag{7.3}$$

is an appropriate solution. The two potentials will be designated by

V_+ for $z < 0$, above the mass plane
V_- for $z > 0$, below the mass plane

respectively.

The Earth's surface is at $z = -d$ and the gravity value on it is the downward derivative of V_+:

$$g(x, -d) = \left|\frac{\partial V_+}{\partial z}\right|_{z=-d} = \sum_m mA_m \begin{matrix}\cos\\ \sin\end{matrix} mx e^{-md}. \tag{7.4}$$

Now we have to correlate this gravity distribution with the underground mass distribution.

It has been assumed that the underground mass is distributed on one single horizontal plane. This plane distribution of condensed mass is a good approximation in a number of actual problems in geophysics. If, for instance, two rock layers having different densities, ρ and ρ', are bounded by an interface having an undulation amplitude $h(x)$, the mass distribution may be approximated by a plane distribution of mass $(\rho' - \rho)h(x)$ condensed on the average depth of the interface, as far as the resulting gravity is concerned. Let this mass distribution condensed on the plane $z = 0$ be expressed by

$$M_{(x,0)} = \sum_m C_m \begin{matrix}\cos\\ \sin\end{matrix} mx. \tag{7.5}$$

We now apply Gauss's theorem to a thin rectangle shown in Figure 7.1. According to the theorem,

$$\left|\frac{\partial V_+}{\partial z} - \frac{\partial V_-}{\partial z}\right|_{z=0} \Delta x = \left|\sum_m mA_m \begin{matrix}\cos\\ \sin\end{matrix} mx e^{mz} - \sum_m mA_m \begin{matrix}\cos\\ \sin\end{matrix} mx e^{-mz}\right|_{z=0} \Delta x$$

$$= 2\sum_m mA_m \begin{matrix}\cos\\ \sin\end{matrix} mx \cdot \Delta x \tag{7.6}$$

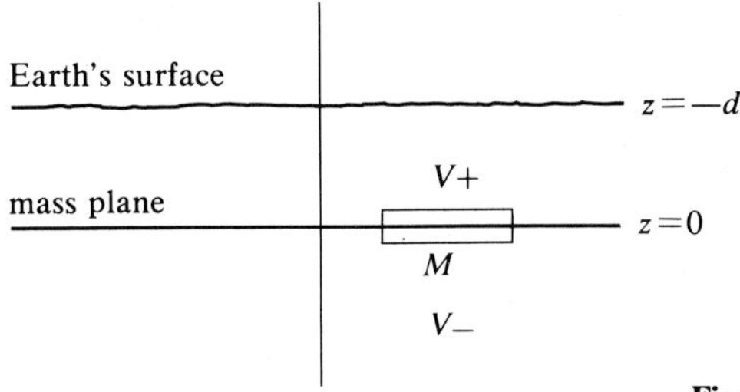

Figure 7.1 A mass plane at a depth d.

must be equal to $4\pi G$ times the mass within the thin rectangle. Therefore

$$2mA_m = 4\pi G C_m$$

or

$$A_m = \frac{2\pi G}{m} C_m.$$

On the Earth's surface $z = -d$,

$$g_{(x,-d)} = \left| \frac{\partial V_+}{\partial z} \right|_{z=-d} = \sum_m mA_m \frac{\cos}{\sin} mx \mathrm{e}^{-md},$$

and let this be expressed by

$$g_{(x,-d)} = \sum_m B_m \frac{\cos}{\sin} mx,$$

where

$$B_m = mA_m \mathrm{e}^{-md}.$$

From these relations, it is seen that

$$A_m = \frac{B_m}{m} \mathrm{e}^{md} = \frac{2\pi G}{m} C_m, \tag{7.7}$$

$$B_m = mA_m \mathrm{e}^{-md} = 2\pi G \mathrm{e}^{-md} C_m, \tag{7.8}$$

$$C_m = \frac{m}{2\pi G} A_m = \frac{B_m}{2\pi G} \mathrm{e}^{md} \tag{7.9}$$

and that, if the observed surface gravity distribution is expressed by

$$g_{(x,\,-d)} = \sum_m B_m \begin{matrix}\cos\\ \sin\end{matrix}\, mx,$$

the responsible mass distribution at the depth d is given by

$$M_{(x,\,0)} = \frac{1}{2\pi G} \sum_m B_m \begin{matrix}\cos\\ \sin\end{matrix}\, mx\,\mathrm{e}^{md}. \tag{7.10}$$

This relation makes it possible for us to calculate the mass distribution directly from the observed surface gravity distribution. Notice that a factor e^{md} comes into the expression for $M(x)$. Even if the amplitude of gravity variation B_m on the Earth's surface is the same, the corresponding

$$C_m = \frac{B_m}{2\pi G}\,\mathrm{e}^{md}$$

will be larger for larger m (shorter wavelength) and deeper d. Figure 7.2 illustrates this. The curves in the figure represent

$$y = \sin x\,\mathrm{e}^{d} + \sin 3x\,\mathrm{e}^{3d}$$

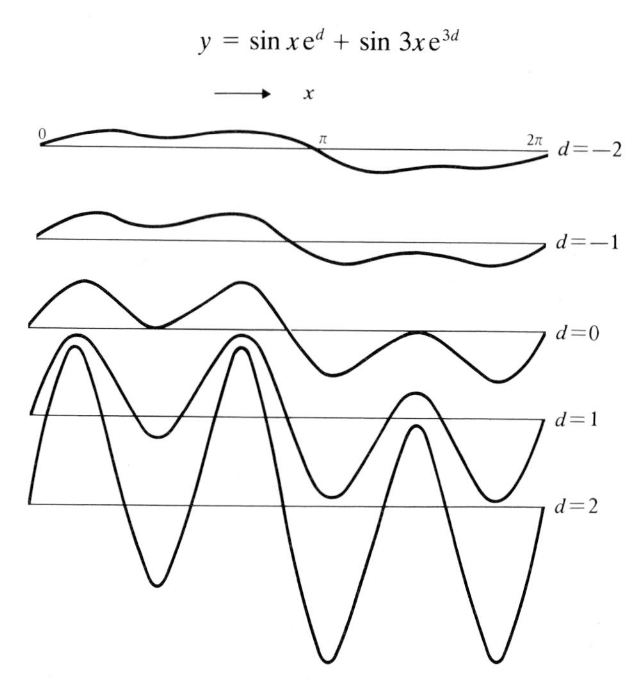

Figure 7.2 $y = \sin x\ \mathrm{e}^{d} + \sin 3x\ \mathrm{e}^{3d}$ for $d = -2, -1, 0, 1$ and 2.

for five values of $d = -2, -1, 0, 1$, and 2. It is clearly seen that when d is increased, the shorter wavelength component $\sin 3x e^{3d}$ grows much faster in amplitude than the longer wavelength component $\sin x e^{d}$.

Finally, it should be noticed that the value of d in the above formula is expressed in radian when the domain of interest is taken to be 2π. If the domain is actually D in length, the actual depth δ corresponding to d is

$$\delta = \frac{D}{2\pi} d.$$

7.2 Fourier series

In the above discussion, the expression

$$\begin{aligned} y = f(x) &= \sum_m \alpha_m \begin{matrix}\cos\\ \sin\end{matrix} mx \\ &= a_0 + a_1 \cos x + a_2 \cos 2x + a_3 \cos 3x + \cdots \\ &\quad + b_1 \sin x + b_2 \sin 2x + b_3 \sin 3x + \cdots \end{aligned} \tag{7.11}$$

has frequently been used. This is called a Fourier series (J. B. J. Fourier, 1768–1830) and the a and b are called Fourier coefficients. An ordinary continuous one-valued function $f(x)$ given within a domain $0 \leqslant x \leqslant 2\pi$ can be expressed by the sum of trigonometrical functions having various amplitudes a_m and b_m with mx as arguments ($m = 0, 1, 2 \ldots$). If the values of $f(x)$ are known continuously throught $0 \leqslant x \leqslant 2\pi$, then the Fourier coefficients a_m and b_m are as follows;

$$a_0 = \frac{1}{2\pi} \int_0^{2\pi} f(x)\, dx,$$

$$a_m = \frac{1}{\pi} \int_0^{2\pi} f(x) \cos mx\, dx,$$

$$b_m = \frac{1}{\pi} \int_0^{2\pi} f(x) \sin mx\, dx.$$

If the values of $y = f(x)$ are known only at $2m$ equidistant points within $0 \leqslant x \leqslant 2\pi$, it is possible to find the values of the Fourier coefficients, $2m$ in number, such that they will make the values of the series equal to the given $2m$ values of y.

In a simple case where the values of y are known to be $y_0, y_1 \ldots y_5$ ($y_6 = y_0$) at six equidistant points within $0 \leqslant x \leqslant 2\pi$, the six Fourier coefficients $a_0, a_1, a_2, a_3, b_1, b_2$ are as follows:

$$a_0 = \tfrac{1}{6}(y_0 + y_1 + y_2 + y_3 + y_4 + y_5),$$

$$a_1 = \tfrac{1}{3}(y_0 \cos 0^\circ + y_1 \cos 60^\circ + y_2 \cos 120^\circ + y_3 \cos 180^\circ + y_4 \cos 240^\circ + y_5 \cos 300^\circ),$$

$$a_2 = \tfrac{1}{3}(y_0 \cos 0^\circ + y_1 \cos 120^\circ + y_2 \cos 240^\circ + y_3 \cos 360^\circ + y_4 \cos 480^\circ + y_5 \cos 600^\circ),$$

$$a_3 = \tfrac{1}{6}(y_0 \cos 0^\circ + y_1 \cos 180^\circ + y_2 \cos 360^\circ + y_3 \cos 540^\circ + y_4 \cos 720^\circ + y_5 \cos 900^\circ),$$

$$b_1 = \tfrac{1}{3}(y_0 \sin 0^\circ + y_1 \sin 60^\circ + y_2 \sin 120^\circ + y_3 \sin 180^\circ + y_4 \sin 240^\circ + y_5 \sin 300^\circ),$$

$$b_2 = \tfrac{1}{3}(y_0 \sin 0^\circ + y_1 \sin 120^\circ + y_2 \sin 240^\circ + y_3 \sin 360^\circ + y_4 \sin 480^\circ + y_5 \sin 600^\circ),$$

or putting the numerical values in the sine and cosine terms, the six Fourier coefficients are:

$$a_1 = \tfrac{1}{6}(y_0 + y_1 + y_2 + y_3 + y_4 + y_5),$$

$$a_1 = \tfrac{1}{3}(y_0 + \tfrac{1}{2}y_1 - \tfrac{1}{2}y_2 - y_3 - \tfrac{1}{2}y_4 + \tfrac{1}{2}y_5),$$

$$a_2 = \tfrac{1}{3}(y_0 - \tfrac{1}{2}y_1 - \tfrac{1}{2}y_2 + y_3 - \tfrac{1}{2}y_4 - \tfrac{1}{2}y_5),$$

$$a_3 = \tfrac{1}{6}(y_0 - y_1 + y_2 - y_3 + y_4 - y_5),$$

$$b_1 = \tfrac{1}{3}(\tfrac{\sqrt{3}}{2}y_1 + \tfrac{\sqrt{3}}{2}y_2 - \tfrac{\sqrt{3}}{2}y_4 - \tfrac{\sqrt{3}}{2}y_5),$$

$$b_2 = \tfrac{1}{3}(\tfrac{\sqrt{3}}{2}y_1 - \tfrac{\sqrt{3}}{2}y_2 + \tfrac{\sqrt{3}}{2}y_4 - \tfrac{\sqrt{3}}{2}y_5).$$

To find the underground-mass distribution, we have to evaluate the series

$$M(x) = \frac{1}{2\pi G} \sum_m (a_m e^{md} \cos mx + b_m e^{md} \sin mz) \qquad (7.12)$$

at $2m$ equidistant points. The process of calculating the values of a Fourier series with given coefficients at successive values of x is called

Fourier synthesis. All the calculations needed for Fourier analysis and synthesis can be done manually with convenient aids designed for the purpose, or much faster by using electronic computers. To summarize, the process of gravity interpretation by means of the Fourier series method is as follows:

(a) Analyse a given $g(x)$ into a Fourier series (that is, find the coefficients a_m and b_m) $2m$ in number, such that the values of the series will be equal to $2m$ observed values of $g(x)$ within $0 \leqslant x \leqslant 2\pi$:

$$g(x) = \sum_m (a_m \cos mx + b_m \sin mx).$$

(b) Calculate $a_m e^{md}$ and $b_m e^{md}$ for $m = 0, 1, 2 \ldots m$ with an assumed depth d.

(c) Synthesize the series

$$M(x) = \frac{1}{2\pi G} \sum_m (a_m e^{md} \cos mx + b_m e^{md} \sin mx)$$

for finding $M(x)$.

(d) If $M(x)$ is considered to be caused by an undulation $h(x)$ of the interface of two rock layers having different densities, ρ and ρ', $h(x)$ is found by dividing $M(x)$ by $(\rho' - \rho)$.

7.3 Depth of the mass plane

In many actual geophysical problems, the depth of the mass plane can be approximately judged from geological information. The depth can also be estimated mathematically by the following two considerations.

The Fourier coefficients $a_m e^{md}$ and $b_m e^{md}$ in the expression which gives the mass distribution become larger for large depths and for larger m according to the factor e^{md}. In order that the coefficients will not be unreasonably large, there must be a certain limit for the depth that it cannot exceed.

Suppose a surface distribution of gravity is given by

$$g(x) = B_m \cos mx.$$

The plane mass at a depth d that will produce this gravity distribution is

$$M(x) = \frac{B_m}{2\pi G} e^{md} \cos mx.$$

If this $M(x)$ is considered to be produced by an undulation of the interface between two rock layers having different densities ρ and ρ', the amplitude $h(x)$ of the undulation of the interface is given by

$$h(x) = \frac{B_m e^{md}}{2\pi G(\rho' - \rho)} \cos mx \qquad (7.13)$$

If this h becomes larger than d, it will mean that the interface comes out of the Earth's surface, which is absurd. In order that this will not occur, there must be a condition that

$$h \leqslant d$$

or

$$\log_e \frac{B_m}{2\pi G(\rho' - \rho)} + md \leqslant \log_e d.$$

From this, it is seen that if, for instance, $\rho' - \rho = 0.3$, $B_m = 10$ mgal and the wavelength = 20 km, the average depth of the interface cannot be larger than 5 km.

Another way to estimate the limiting depth of the mass plane is based on a consideration of the spectrum-amplitude distribution of $B_m e^{md}$ according to m. If the spectrum-amplitude becomes larger with larger m, the series $M(x)$ will not converge smoothly and this must be avoided.

The following is an example of estimating the limiting depth of the mass plane by this consideration. The example refers to an east–west cross-section of the northern part of Honshu, the main island of Japan, where the gravity and topography can be regarded as varying in the east–west direction only. The average east–west profile of gravity distribution in this area was analyzed in a Fourier series up to $m = 18$, taking 1170 km as 2π. The coefficients $B_m e^{md}$ calculated for $d = 0.1$, 0.2 and 0.3 are plotted in Figure 7.3 according to m. Since the length of 1170 km was taken as 2π, $d = 0.1$ corresponds to

$$\frac{1170}{2\pi} \times 0.1 = 18.6 \text{ km}$$

in actual length. For $d = 0.1$, the coefficient $B_m e^{md}$ decreases smoothly with m. The case $d = 0.3$ is to be rejected because the coefficient increases with m unnaturally. The case $d = 0.2$ is not quite satisfactory either. The conclusion is therefore that the depth of the mass plane in

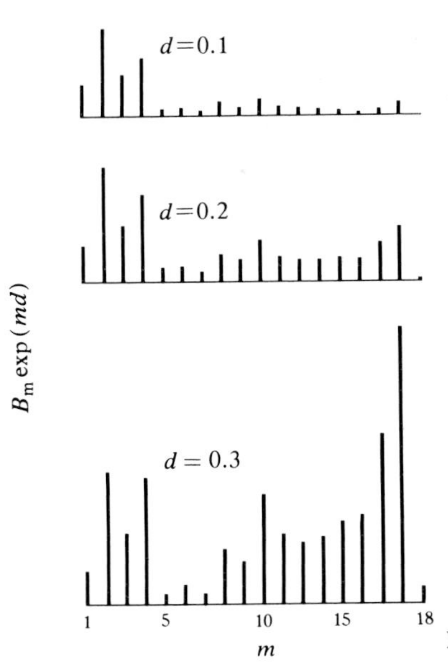

Figure 7.3 $B_m e^{md}$ for $d = 0.1, 0.2$ and 0.3.

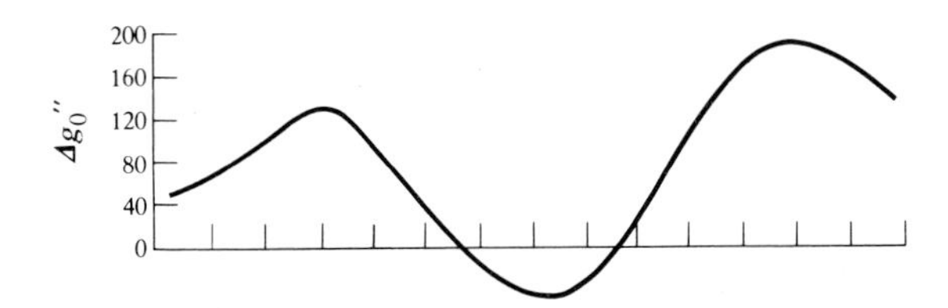

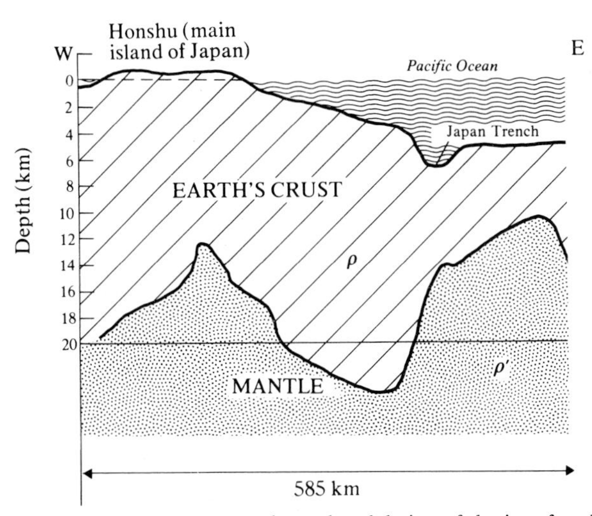

Figure 7.4 Bouguer anomaly, topography and undulation of the interface between the Earth's crust and mantle across the northern part of Honshu, the main island of Japan.

this area cannot be larger than $d = 0.2 = 37.2$ km. Figure 7.4 shows the east–west profiles of gravity, topography and the undulation of the interface between the Earth's crust and the mantle calculated on the assumption that $d = 20\text{ km} = 0.107$ and $\rho' - \rho = 0.3$.

7.4 Assumption of plane distribution of mass

So far, a mass distribution produced by an undulation $h(x)$ of the interface between two rock layers having different densities ρ' and ρ has been replaced by the mass $h(\rho' - \rho)$ condensed on the average depth of the interface. This is a good approximation as can be seen by the following simple example. If there is a plane mass

$$M(x) = C_m \cos mx$$

at a depth d, the resulting gravity distribution on the Earth's surface is

$$g(x) = 2\pi G C_m \mathrm{e}^{-md} \cos mx.$$

If this same mass is considered to be distributed uniformly in the vertical direction between two depths a and b, the resulting gravity distribution on the Earth's surface will be

$$g(x) = 2\pi G \frac{C_m}{b - a} \cos mx \int_a^b \mathrm{e}^{-mz}\, \mathrm{d}z. \tag{7.14}$$

But

$$\int_a^b \mathrm{e}^{-mz}\, \mathrm{d}z = \frac{1}{m}(\mathrm{e}^{-ma} - \mathrm{e}^{-mb}).$$

If ma and mb are small compared with 1, the integral

$$\int_a^b \mathrm{e}^{-mz}\, \mathrm{d}z$$

reduces to

$$\frac{1}{m}(\mathrm{e}^{-ma} - \mathrm{e}^{mb}) = \frac{1}{m}\left(1 - ma + \frac{m^2a^2}{2} - 1 + mb - \frac{m^2b^2}{2}\right)$$

$$\approx (b - a)\left\{1 - \frac{m}{2}(b + a)\right\}$$

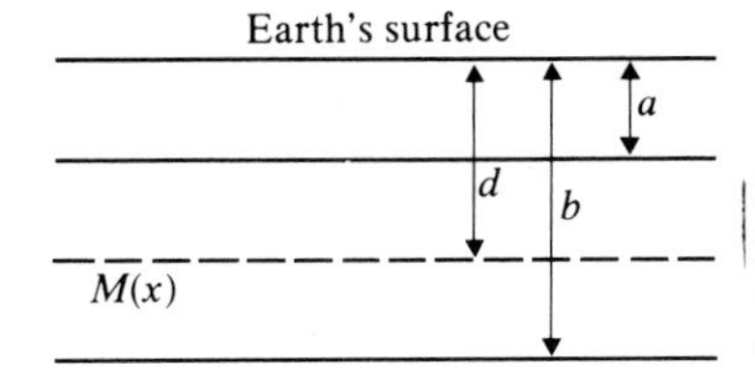

Figure 7.5 A condensed mass $M(x)$ on a single plane at a depth d and a uniformly distributed mass $M(x)$ in the vertical direction between two depths a and b.

and the resulting gravity will be

$$g(x) = 2\pi G C_m \cos mx \left\{ 1 - \frac{m}{2}(b + a) \right\}. \tag{7.15}$$

Comparing this with the distribution in the case of a condensed mass

$$g(x) = 2\pi G C_m \cos mx \, e^{-md}$$

$$\approx 2\pi G C_m \cos mx(1 - md),$$

it is seen that if

$$d = \tfrac{1}{2}(b + a)$$

the two gravity distributions are the same. This means that the underground mass distributed uniformly between two depths a and b in the vertical direction can be approximated by a plane mass condensed at their middle depth as far as the resulting gravity is concerned.

7.5 The (sin x)/x method

As was seen in the preceding sections, the Fourier series method is very useful for finding underground mass distributions directly from an observed distribution of gravity on the Earth's surface. The method, however, is not without its weak points. To express a distribution of a certain quantity within a domain of 2π by a Fourier series implies that the same distribution is repeated infinitely into both sides of the domain (Fig. 7.6a–b). This is not true in actual cases and, for that reason, errors are liable to occur in the results of calculations, especially in the neighbourhood of the boundaries at $x = 0$ and $x = 2\pi$.

Several methods have been proposed for avoiding these. Sato (1954) proposed to insert blank domains as shown in Figure 7.6c, while Dix (1964) proposed to extend the end values to infinity as shown in Figure 7.6d. They derived formulae for gravity interpretations to be used in the respective cases.

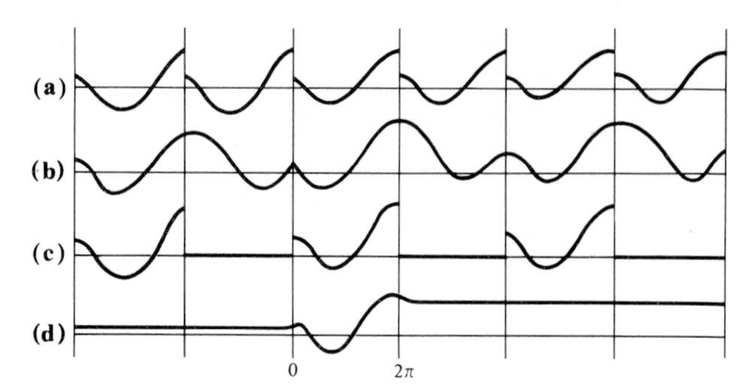

Figure 7.6 Continuation of a Fourier domain: (a) and (b) where the same distribution is repeated infinitely into both sides of the domain; (c) where blank domains are inserted; and (d) where the end values are extended to infinity.

Tomoda and Aki (1955) devised an ingenious application of the function (sin z)/x for gravity interpretations, which could minimize the above-mentioned difficulties at the boundaries.

This function takes the value

$$\frac{\sin x}{x} = 1 \text{ at } x = 0$$

and

$$\frac{\sin x}{x} = 0 \text{ at } x = n\pi$$

(where n is an integer) as shown in Figure 7.7. Also

$$\frac{\sin x}{x} = \int_0^1 \cos mx \, dm.$$

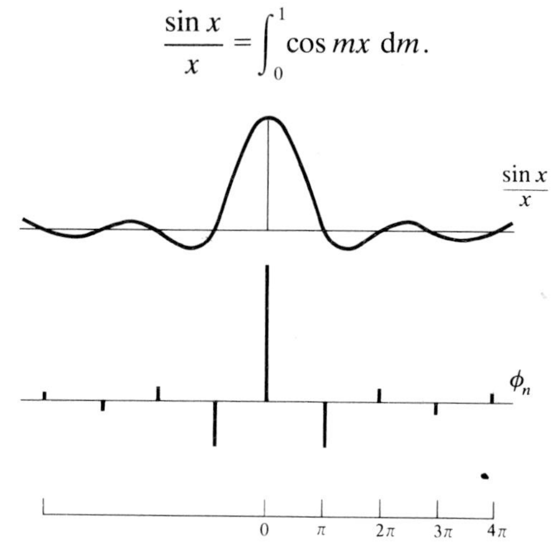

Figure 7.7 Application of the functions $(\sin x)/x$ and ϕ_n for gravity interpretation.

In words, the function $(\sin x)/x$ is the integrated sum of $\cos mx$ with respect to m, all having a unit amplitude and m values continuing from zero to one. No component having a wavelength shorter than 2π is included in it. This function is suitable, therefore, for expressing the distribution of a quantity such that it is equal to 1 at $x = 0$ and is zero at all other grid points which are at distances of integral multiple of π from $x = 0$.

Now if there is a surface gravity distribution such that it is g_p at the pth grid point ($x = 0$) and zero at all other grid points $n\pi$, the distribution can be expressed by

$$g(x) = g_p \frac{\sin x}{x} = g_p \int_0^1 \cos mx \, \mathrm{d}m. \tag{7.16}$$

The mass distribution $M(x)$ at a depth d that will produce this gravity distribution is given by

$$M(x) = \frac{g_p}{2\pi G} \int_0^1 \cos mx \, \mathrm{e}^{md} \, \mathrm{d}m.$$

The integral on the right-hand side of this expression is

$$\phi_x(d) = \int_0^1 \cos mx \, \mathrm{e}^{md} \, \mathrm{d}m$$

$$= \frac{d}{d^2 + x^2} (\mathrm{e}^d \cos x - 1) + \frac{x \mathrm{e}^d}{d^2 + x^2} \sin x$$

and at $x = \pm n\pi$

$$\phi_n(d) = \frac{d}{d^2 + (n\pi)^2} (\pm \mathrm{e}^d - 1)$$

($+ \mathrm{e}^d$ when n is even, and $-\mathrm{e}^d$ when n is odd). If the depth d of the mass plane is equal to the distance between two consecutive grid points, d is π, and if d is one-half of the grid-point distance, d is $\pi/2$, so that

$$\phi_n(\pi) = \frac{1}{(n^2 + 1)\pi} (\pm \mathrm{e}^\pi - 1)$$

$$\phi_n\left(\frac{\pi}{2}\right) = \frac{2}{(4n^2 + 1)\pi} (\pm \mathrm{e}^{\pi/2} - 1).$$

The numerical values of $\phi_n(\pi)$ and $\phi_n(\pi/2)$ are given in Table 7.1.

Table 7.1 Values of $\phi_n(\pi)$ and $\phi_n(\pi/2)$.

$\pm n$	$\phi_n(\pi)$	$\phi_n(\pi/2)$
0	7.056	2.425
1	−3.843	−0.739
2	1.410	0.143
3	−0.768	−0.100
4	0.414	0.037
5	−0.295	−0.036
6	0.190	0.017
7	−0.155	−0.019
8	0.168	0.010
9	−0.094	−0.012
10	0.072	0.006
⋮	...	...

It is to be noted that the values of ϕ_n are alternately positive and negative. The mass beneath the $(p \pm n)$th grid point needed for producing g_p at the pth grid point and zero gravity at all other grid points is

$$m_{p\pm n} = \frac{g_p}{2\pi G}\phi_n.$$

Since any gravity distribution

$$\ldots g_{p-3}, g_{p-2}, g_{p-1}, g_p, g_{p+1}, g_{p+2}, g_{p+3}, \ldots$$

can be regarded as a superposition of the following distributions,

$$\begin{array}{ccccccccc}
\ldots & g_{p-3}, & 0, & 0, & 0, & 0, & 0, & 0, & \ldots \\
\ldots & 0, & g_{p-2}, & 0, & 0, & 0, & 0, & 0, & \ldots \\
\ldots & 0, & 0, & g_{p-1}, & 0, & 0, & 0, & 0, & \ldots \\
\ldots & 0, & 0, & 0, & g_p, & 0, & 0, & 0, & \ldots \\
\ldots & 0, & 0, & 0, & 0, & g_{p+1}, & 0, & 0, & \ldots \\
\ldots & 0, & 0, & 0, & 0, & 0, & g_{p+2}, & 0, & \ldots \\
\ldots & 0, & 0, & 0, & 0, & 0, & 0, & g_{p+3}, & \ldots
\end{array}$$

the underground mass M_{p+n} beneath the $(p+n)$th grid point that is needed for producing the distribution

$$\ldots g_{p-3}, g_{p-2}, g_{p-1}, g_p, g_{p+1}, g_{p+2}, g_{p+3}, \ldots \tag{7.17}$$

is given by

$$M_{p+n} = \frac{1}{2\pi G}(\ldots g_{p+n-3}\phi_3 + g_{p+n-2}\phi_2 + g_{p+n-1}\phi_1 + g_{p+n}\phi_0 + g_{p+n+1}\phi_1 + g_{p+n+2}\phi_2 + g_{p+n+3}\phi_3 + \ldots).$$

By changing p by one successively, the masses beneath all the grid points can be calculated by this formula. The calculations are easy by using a suitable multiplication table.

This method is very neat, but it must be added that there still remains a problem of deciding how far we should go to distant grid points.

7.6 Finite weight function method

The difficulties encountered in using a Fourier series and the $(\sin x)/x$ function were pointed out. They can be formally overcome by limiting the number of g values to be used to a finite small number and by calculating the mass beneath its central point only, instead of trying to find the underground masses beneath every gravity point within the extent of 2π.

We will limit the values of gravity to seven equidistant points only, for instance. If this extent is taken as 2π and if the gravity values at the seven points are

$$g_{p-3}, g_{p-2}, g_{p-1}, g_p, g_{p+1}, g_{p+2}, g_{p+3},$$

this gravity distribution is given by a Fourier series

$$g(x) = a_0 + a_1 \cos x + a_2 \cos 2x + a_3 \cos 3x + b_1 \sin x + b_2 \sin 2x, \tag{7.18}$$

where

$$a_0 = \tfrac{1}{6}(\tfrac{1}{2}g_{p-3} + g_{p-2} + g_{p-1} + g_p + g_{p+1} + g_{p+2} + \tfrac{1}{2}g_{p+3}),$$

$$a_1 = \tfrac{1}{3}(\tfrac{1}{2}g_{p-3} + \tfrac{1}{2}g_{p-2} - \tfrac{1}{2}g_{p-1} - g_p + \tfrac{1}{2}g_{p+1} + \tfrac{1}{2}g_{p+2} + \tfrac{1}{2}g_{p+3})$$

$$a_2 = \tfrac{1}{3}(\tfrac{1}{2}g_{p-3} - \tfrac{1}{2}g_{p-2} - \tfrac{1}{2}g_{p-1} + g_p - \tfrac{1}{2}g_{p+1} - \tfrac{1}{2}g_{p+2} + \tfrac{1}{2}g_{p+3}),$$

$$a_3 = \tfrac{1}{6}(\tfrac{1}{2}g_{p-3} - g_{p-2} + g_{p-1} - g_p + g_{p+1} - g_{p+2} + \tfrac{1}{2}g_{p+3}).$$

b_1 and b_2 will not be needed in subsequent calculations. The mass M beneath the central point of the domain is given by putting $x = \pi$ in the expression:

$$M(\pi) = \frac{1}{2\pi G}(a_0 + a_1 e^d \cos x + a_2 e^{2d} \cos 2x + a_3 e^{3d} \cos 3x$$

$$+ b_1 e^d \sin x + b_2 e^{2d} \sin 2x),$$

that is

$$M(\pi) = \frac{1}{2\pi G}(a_0 - a_1 e^d + a_2 e^{2d} - a_3 e^{3d}). \tag{7.19}$$

$\sin x$ and $\sin 2x$ are both zero at $x = \pi$.

By using a_0, a_1, a_2 and a_3 values given above, it is seen that

$$M(\pi) = \frac{1}{2\pi G}\left\{g_{p-3}\left(\frac{1}{12} - \frac{e^d}{6} + \frac{e^{2d}}{6} - \frac{e^{3d}}{12}\right)\right.$$

$$+ g_{p-2}\left(\frac{1}{6} - \frac{e^d}{6} - \frac{e^{2d}}{6} + \frac{e^{3d}}{6}\right)$$

$$+ g_{p-1}\left(\frac{1}{6} + \frac{e^d}{6} - \frac{e^{2d}}{6} - \frac{e^{3d}}{6}\right)$$

$$+ g_p\left(\frac{1}{6} + \frac{e^d}{3} + \frac{e^{2d}}{3} + \frac{e^{3d}}{6}\right)$$

$$+ g_{p+1}\left(\frac{1}{6} + \frac{e^d}{6} - \frac{e^{2d}}{6} - \frac{e^{3d}}{6}\right)$$

$$+ g_{p+2}\left(\frac{1}{6} - \frac{e^d}{6} - \frac{e^{2d}}{6} + \frac{e^{3d}}{6}\right)$$

$$\left.+ g_{p+3}\left(\frac{1}{12} - \frac{e^d}{6} + \frac{e^{2d}}{6} - \frac{e^{3d}}{12}\right)\right\}$$

$$= \frac{1}{2\pi G}(g_{p-3}\phi_3 + g_{p-2}\phi_2 + g_{p-1}\phi_1 + g_p\phi_0$$

$$+ g_{p+1}\phi_1 + g_{p+2}\phi_2 + g_{p+3}\phi_3).$$

The expressions in the parentheses are numerical constants ϕ_n, if the value of d is fixed. If $2n = 6$ and if the depth of the underground-mass plane is equal to the distance between two consecutive grid points, then

$d = \pi/3$, and if the depth of the mass plane is a half of the grid-point distance, $d = \pi/6$ respectively. The values of the coefficients ϕ_n for $d = \pi/3$ and $d = \pi/6$ are given in Table 7.2.

Table 7.2 Values of ϕ_n.

$\pm n$	$\phi_n(\pi/3)$	$\phi_n(\pi/6)$
0	7.679	2.484
1	−4.567	−0.829
2	2.196	0.212
3	−0.967	−0.124

The mass M_p beneath the pth grid point can be calculated according to

$$M_p = \frac{1}{2\pi G} \sum_{n=-3}^{n=3} g_{p+n} \phi_n, \tag{7.20}$$

and by changing p by one successively, the masses beneath all the grid points can be calculated. In the above example, seven gravity points were taken into consideration. The number can be increased if desired, but in doing so, it is inevitable that the underground mass will be known at fewer points.

7.7 Two-dimensional cases

So far, various methods for gravity interpretations in one-dimensional cases have been explained. These methods can be extended to be applicable to two-dimensional cases. In these cases, a double Fourier series is to be used instead of a single Fourier series.

A double Fourier series has a form:

$$f(xy) = \sum_m \sum_n \alpha_{mn} \begin{matrix}\cos\\ \sin\end{matrix}\, mx \begin{matrix}\cos\\ \sin\end{matrix}\, ny. \tag{7.21}$$

Suppose a distribution of a certain quantity $f(xy)$ is given within a square, $0 \leqslant x \leqslant 2\pi$, $0 \leqslant y \leqslant 2\pi$. The distribution of the value of $f(xy)$ in the direction of x along a certain value of y can be expressed by a single Fourier series of x such as

$$f(xy) = \sum_m \beta_m(y) \begin{matrix}\cos\\ \sin\end{matrix}\, mx.$$

The coefficient β_m changes according to y, so that it can be expressed by a single Fourier series of y such as

$$\beta_m(y) = \sum_n \gamma_n \begin{matrix}\cos\\ \sin\end{matrix} ny.$$

Then as a whole, $f(xy)$ is

$$f(xy) = \sum_m \sum_n \beta_m \gamma_n \begin{matrix}\cos\\ \sin\end{matrix} mx \begin{matrix}\cos\\ \sin\end{matrix} ny.$$

If $\beta_m \gamma_n$ is written as α_{mn}, then

$$f(xy) = \sum_m \sum_n \alpha_{mn} \begin{matrix}\cos\\ \sin\end{matrix} mx \begin{matrix}\cos\\ \sin\end{matrix} ny.$$

Using this double Fourier series, if a distribution of gravity values $g(xy)$ is

$$g(xy) = \sum_m \sum_n B_{mn} \begin{matrix}\cos\\ \sin\end{matrix} mx \begin{matrix}\cos\\ \sin\end{matrix} ny,$$

then the underground mass M at a depth d that will produce this $g(xy)$ is given by

$$M(xy) = \frac{1}{2\pi G} \sum_m \sum_n B_{mn} \exp\{\sqrt{(m^2 + n^2)}d\} \begin{matrix}\cos\\ \sin\end{matrix} mx \begin{matrix}\cos\\ \sin\end{matrix} ny. \quad (7.22)$$

In the two-dimensional cases, the coefficients are $4 \times m \times n$ in number. For instance, if 2π is divided into six both in x- and y-directions, the number of coefficients will be $4 \times 6 \times 6 = 144$, instead of 6 in one-dimensional case. If 2π is divided into 18, the number of coefficients needed are as many as $4 \times 18 \times 18 = 1296$. The factor 4 is needed because there are four different combinations of $\cos mx \cos ny$, $\cos mx \sin ny$, $\sin mx \cos ny$ and $\sin mx \sin ny$.

The $(\sin x)/x$ method can also be extended to be applicable to two-dimensional cases. In these cases, the integral

$$\phi' = \int_0^1 \int_0^1 \cos mx \cos ny \exp\{\sqrt{(m^2 + n^2)}\} \, \mathrm{d}m \, \mathrm{d}n \quad (7.23)$$

has to be evaluated, corresponding to

$$\phi = \int_0^1 \cos mx \exp(md) \, dm$$

in one-dimensional cases. This integral ϕ' cannot be evaluated analytically and we have to resort to numerical calculations. There are several tables which give the numerical values of ϕ' (Tsuboi *et al.* 1958, Kanamori 1963, Takeuchi & Sato 1964) for various values of d. The last one is the most comprehensive. Table 7.3 gives the values of ϕ' for $d = \pi/2$ (a half of the grid distance) taken from the last table.

Table 7.3 Values of $\phi'(\pi/2)$.

$n \backslash m$	−4	−3	−2	−1	0	1	2	3	4
−4	...	...	...	...	...	...	...	...	...
−3	...	−0.000	−0.003	0.003	−0.095	0.003	−0.003	−0.000	...
−2	...	−0.003	0.000	−0.020	0.192	−0.020	0.000	−0.003	...
−1	...	0.003	−0.020	0.022	−0.742	0.022	−0.020	0.003	...
0	...	−0.095	0.192	−0.742	3.657	−0.742	0.192	−0.095	...
1	...	0.003	−0.020	0.022	−0.742	0.022	−0.020	0.003	...
2	...	−0.003	0.000	−0.020	0.192	−0.020	0.000	−0.003	...
3	...	−0.000	−0.003	0.003	−0.095	0.003	−0.003	−0.000	...
4	...	...	...	...	...	...	...	...	...

To apply this method to two-dimensional gravity interpretations, gravity values g_{mn} at square grid points and ϕ'_{mn} values at the corresponding points are multiplied and the products $g_{mn}\phi'_{mn}$ at all the grid points added. When the sum of all the products is divided by $2\pi G$, this will give the mass right beneath the point corresponding to $m = n = 0$. Then we shift the ϕ' value mesh step by step relative to the g value mesh and do the same calculations. In this way, the underground-mass distribution at a depth d will be found.

To what extent m and n should go in the calculations is a matter which is to be decided in each particular case. If such a decision is not possible, it is perhaps wise to limit the extent of m and n mechanically from the outset, in the same way as is done in the weight-function method described in Section 7.6. In a case where m and n are limited to ± 2, for instance, the weight function ϕ' for $d = \pi/4$ (a half of the grid distance) has the values as given in Table 7.4. The sum of the products of g_{mn} and ϕ'_{mn} at 5×5 grid points when divided by $2\pi G$ gives the mass right beneath the central point $x = 0$, $y = 0$.

Table 7.4 Values of $\phi'(\pi/4)$.

$n \backslash m$	−2	−1	0	1	2
−2	0.001	−0.032	0.241	−0.032	0.001
−1	−0.032	0.038	−0.963	0.038	−0.032
0	0.241	−0.963	3.995	−0.963	0.241
1	−0.032	0.038	−0.963	0.038	−0.032
2	0.001	−0.032	0.241	−0.032	0.001

The change in the values of ϕ' in Tables 7.3 and 7.4 differs considerably according to their directions from the origin. Although this is a necessary consequence which arises from the method of taking x- and y-axes, this is undesirable from the physical point of view. It is therefore better to use cylindrical co-ordinates in these two-dimensional problems. This will be the subject of Chapter 9.

References and further reading

Bullard, B. C. and I. B. Cooper 1938. The determination of the masses necessary to produce a given gravity field. *Proc. R. Soc., Lond.* **194**, 273.

Dix, C. H. 1964. Use of the function $(\sin x)/x$ in gravity problems. *Proc. Jap. Acad.* **40**, 276.

Kanamori, H. 1963. A new method for downward continuation of two-dimensional gravity distribution. *Proc. Jap. Acad.* **39**, 469.

Rikitake, T. 1952. Analysis of geomagnetic field by use of hermite functions. *Bull. Earthquake Res. Inst.* **30**, 293.

Sato, Y. 1954. A note on Tsuboi-Nagata's method. *Bull. Earthquake Res. Inst.* **32**, 259.

Takeuchi, H. and M. Saito 1964. Numerical tables useful in three-dimensional gravity interpretation. *Bull. Earthquake Res. Inst.* **42**, 39.

Tomoda, T. and K. Aki 1955. Use of the function $(\sin x)/x$ in gravity problems. *Proc. Jap. Acad.* **31**, 443.

Tsuboi, C. 1938. Gravity anomalies and the corresponding subterranean mass distribution. *Proc. Imp. Acad., Japan* **14**, 170.

Tsuboi, C. 1959a. *Applications of Fourier Series for computing gravity anomalies and other gravimetric quantities at any elevation from surface gravity anomalies*. Rep. Inst. Geod. Photogr. Cartogr. Ohio State University, no. 1.

Tsuboi, C. 1959b. *Application of Double Fourier Series to computing gravity anomalies and other gravimetric quantities at higher elevations from surface gravity anomalies*. Rep. Inst. Geod. Photogr. Cartogr. Ohio State University, no. 2.

Tsuboi, C. 1959c. *Application of* $(\sin x)/x$ *and other similar functions to computing gravity anomalies at higher elevation, starting from given surface anomalies.* Rep. Inst. Geod. Photogr. Cartogr. Ohio State University, no. 3.

Tsuboi, C. 1959d. *Weight function method for computing gravity anomalies at higher elevations. Starting from given surface gravity anomalies.* Rep. Inst. Geod. Photogr. Cartogr. Ohio State University, no. 4.

Tsuboi, C. and T. Fuchida 1938. Relation between gravity anomalies and the corresponding subterranean mass distribution (II). *Bull. Earthquake Res. Inst.* **16**, 273.

Tsuboi, C., C. H. G. Oldham and V. B. Waithman 1958. Numerical tables facilitating three-dimensional gravity interpretations. *J. Phys. Earth* **6**, 7.

8 Deflection of the vertical and undulation of the geoid in cartesian co-ordinates

8.1 Deflection of the vertical

In order to avoid confusion, the gravity anomaly will be written as Δg in Chapters 8 and 9, instead of g as in previous chapters. g in this chapter means normal gravity, 980.

If there is an anomalous mass under the Earth's surface as shown in Figure 8.1, it exerts an attraction which works not only in the z-direction but also in the x-direction. This causes a change in the direction of the vertical by an angle θ when compared with the case without the anomalous mass. This angle θ is equal to X/g ($g = 980$) and is called the **deflection of the vertical**. The vertical directions at two points A and B on a geometrically horizontal Earth's surface are parallel in the case when there is no anomalous mass but cannot be parallel any more if there is a mass in their neighbourhood.

If V is the potential due to the anomalous mass,

$$X = \frac{\partial V}{\partial x}$$

then

$$\theta = \frac{X}{g} = \frac{1}{g}\frac{\partial V}{\partial x}.$$

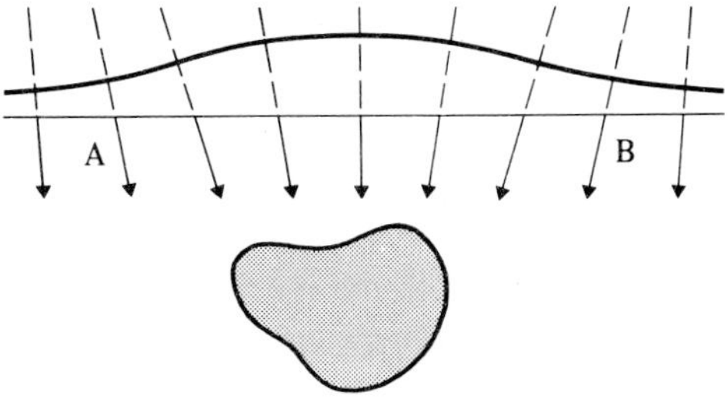

A B

Figure 8.1 Underground mass, direction of gravity forces and undulation of the geoid.

If $X = 10$ mgal, for instance, the angle of deflection of the vertical is

$$\theta = \frac{10^{-2}}{980} \approx 10^{-5} = 2''$$

Since

$$\theta = \frac{1}{g}\frac{\partial V}{\partial x}$$

and

$$\Delta g = \frac{\partial V}{\partial z}$$

are derived from the same potential V, there must be a mathematical relationship between θ and Δg. In this chapter, the relationship between θ and Δg will be discussed in cartesian co-ordinates; more general cases in cylindrical and spherical co-ordinates will be dealt with in Chapters 9 and 10.

8.2 Undulation of the geoid

If the Earth's surface is a geometrical plane and if there is no anomalous mass under it, gravity forces are geometrically parallel everywhere and the geoid will also be a geometrical plane. But if there is an anomalous mass under the Earth's surface, gravity forces will not be parallel so that the shape of the geoid which intersects the direction of gravity forces at right angles cannot be a geometrical plane and will therefore have an undulation. This undulation, however, cannot be recognized by a person who is on this surface and thinks it is horizontal everywhere. In short, the undulation of the geoid means its deviation from the geometrical plane.

8.3 Gravity anomalies and undulation of the geoid

In Figure 8.1, gravity forces are not parallel everywhere. If a curve is drawn such that it intersects the direction of gravity forces at right angles everywhere, the curve will swell upward above the underground positive mass. This is also where Δg is large.

If Δg is large, this means that the vertical gradient of potential V is

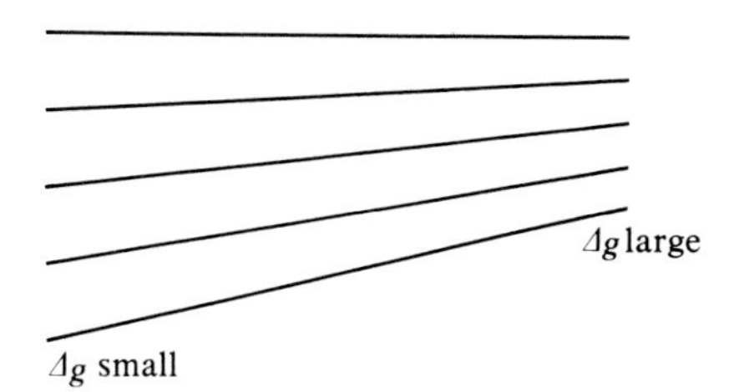

Figure 8.2 Spacing between two consecutive equipotential surfaces is narrower where Δg is large.

large, so that the spacing between two consecutive equipotential surfaces is narrower than elsewhere, as shown in Figure 8.2. At higher elevations, Δg will tend to be uniform and so will the spacing between two consecutive equipotential surfaces. Then, in order to have narrower spacings between two consecutive equipotential surfaces near the Earth's surface, the geoid must swell upwards where Δg is large.

Let us show this by a simple example. A point mass M is assumed to exist at a depth d, as shown in Figure 8.3. The value of Δg at $x = x$ is

$$\Delta g = GM \frac{d}{(d^2 + x^2)^{3/2}} . \tag{8.1}$$

The x-component of attraction due to the mass is

$$X = GM \frac{x}{(d^2 + x^2)^{3/2}}$$

and the deflection of the vertical is

$$\theta = \frac{X}{g} = \frac{GM}{g} \frac{x}{(d^2 + x^2)^{3/2}} .$$

The undulation height h of the geoid is obtained by integrating θ with respect to x from $-\infty$ to x,

$$h(x) = \int_{-\infty}^{x} \theta \, \mathrm{d}x = \frac{GM}{g} \frac{1}{(d^2 + x^2)^{1/2}} .$$

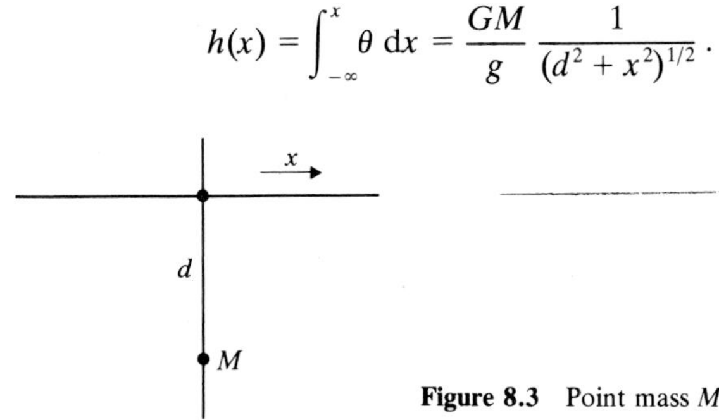

Figure 8.3 Point mass M at depth of d.

At the point $x = 0$ right above the mass

$$\Delta g = \frac{GM}{d^2}$$

$$h = \frac{GM}{g}\frac{1}{d},$$

therefore

$$h = \Delta g \frac{d}{g}$$

or

$$gh = \Delta g \times d.$$

If $d = 10$ km and $\Delta g = 10$ mgal for example,

$$h = 10^{-2} \times 10^{6} \div 980 \approx 10 \text{ cm}.$$

8.4 General cases

Let us discuss this problem in a more general way. The assumptions which are involved are (a) the anomalous mass is on one horizontal plane at a constant depth d, and (b) its surface density varies in the x-direction only. The Fourier series expression of the potential V due to this plane mass is

$$V = \sum_m A_m^{\,c} \cos\frac{2\pi m}{D}x \exp\left(\frac{2\pi m}{D}z\right) + \sum_m A_m^{\,s} \sin\frac{2\pi m}{D}x \exp\left(\frac{2\pi m}{D}z\right). \tag{8.2}$$

Here D is the actual length which was taken as 2π in the Fourier series. On the Earth's surface $z = 0$,

$$\Delta g(x) = \frac{\partial V}{\partial Z}$$

$$= \sum_m \frac{2\pi m}{D} A_m^{\,c} \cos\frac{2\pi m}{D}x + \sum_m \frac{2\pi m}{D} A_m^{\,s} \sin\frac{2\pi m}{D}x, \tag{8.3}$$

$$\theta(x) = \frac{1}{g}\frac{\partial V}{\partial x} = \frac{1}{g}\left(\sum_m -\frac{2\pi m}{D}A_m{}^c \sin\frac{2\pi m}{D}x + \sum_m \frac{2\pi m}{D}A_m{}^s \cos\frac{2\pi m}{D}x\right) \tag{8.4}$$

$$h(x) = \frac{1}{g}\left(\sum_m \int_0^x -\frac{2\pi m}{D}A_m{}^c \sin\frac{2\pi m}{D}x + \sum_m \int_0^x \frac{2\pi m}{D}A_m{}^s \sin\frac{2\pi m}{D}x\right)$$

$$= \frac{1}{g}\sum_m A_m{}^c\left(\cos\frac{2\pi m}{D}x - 1\right) + \frac{1}{g}\sum_m A_m{}^s\left(\sin\frac{2\pi m}{D}x\right). \tag{8.5}$$

Putting

$$\frac{2\pi m}{D}A_m = B_m$$

and taking a particular m, it is seen that

$$\Delta g(x) = B_m{}^c \cos\frac{2\pi m}{D}x + B_m{}^s \sin\frac{2\pi m}{D}x, \tag{8.6}$$

$$\theta(x) = -\frac{B_m{}^c}{g}\sin\frac{2\pi m}{D}x + \frac{B_m{}^s}{g}\cos\frac{2\pi m}{D}x, \tag{8.7}$$

$$h(x) = \frac{D}{2\pi m g}B_m{}^c\left(\cos\frac{2\pi m}{D}x - 1\right) + \frac{D}{2\pi m g}B_m{}^s \sin\frac{2\pi m}{D}x. \tag{8.8}$$

In Δg series and θ series, cosine terms and sine terms are seen to be interchanged, so that there is a phase difference of $\pi/2$ between them. Δg and h are in the same phase but their ratio depends on m. D/m is the wavelength λ_m of this particular component and

$$h_{\max} = \frac{\lambda_m}{2\pi g}B_m.$$

A longer wavelength component will have a larger effect on h. The ratio of amplitudes of Δg and θ does not depend on m, although cosine and sine terms are interchanged. The values of $h_{\max}$ for various λ_m and B_m are given in Table 8.1.

Table 8.1 h_{max} as a function of λ_m and B_m.

$\lambda_m \backslash B_m$	1 mgal	10 mgal	100 mgal
1 km	0.02 cm	0.16 cm	1.62 cm
10 km	0.16 cm	1.62 cm	16.2 cm
100 km	1.62 cm	16.2 cm	162 cm

As an example, the undulation of the geoid across the middle part of Honshu (the main island of Japan) will be calculated. Honshu is about 200 km wide between the Sea of Japan coast and the Pacific Ocean coast, and the Bouguer anomalies decrease from +50 mgal on both coasts to −80 mgal at its central axis. By using the above formula, it is seen that

$$h = \frac{200 \times 10^5}{2\pi \times 980} \times \tfrac{1}{2}(50 + 80) \times 10^{-3}$$

$$\approx 422 \text{ cm.}$$

The geoid is about 4.2 m lower at the central axis of the island than at the coasts.

8.5 Weight function method for finding θ and h from Δg

The weight function method for calculating the anomalous underground mass from Δg was explained before. A similar method can be used for calculating θ from Δg. It has already been shown that if

$$\Delta g = \sum_m B_m{}^c \cos \frac{2\pi m}{D} x + \sum_m B_m{}^s \sin \frac{2\pi m}{D} x,$$

then

$$\theta = \frac{1}{g}\left(- \sum_m B_m{}^c \sin \frac{2\pi m}{D} x + \sum_m B_m{}^s \cos \frac{2\pi m}{D} x \right).$$

D is the actual length which was taken as 2π in the Fourier series. At the middle point $x = D/2$, all the $\sin(2\pi m/D)x$ terms vanish and

$$\cos \frac{2\pi m}{D} x = \pm 1$$

(+1 when m is even, and -1 when m is odd), so that θ at $x = D/2$ becomes simply

$$\theta = \frac{1}{g}(-B_1^s + B_2^s - B_3^s + \cdots). \tag{8.9}$$

If 2π is divided into six intervals, B_1^s *and* B_2^s remain only. And since

$$B_1^s = \frac{1}{3}\left(\frac{\sqrt{3}}{2}\Delta g_1 + \frac{\sqrt{3}}{2}\Delta g_2 - \frac{\sqrt{3}}{2}\Delta g_4 - \frac{\sqrt{3}}{2}\Delta g_5\right)$$

$$B_2^s = \frac{1}{3}\left(\frac{\sqrt{3}}{2}\Delta g_1 - \frac{\sqrt{3}}{2}\Delta g_2 + \frac{\sqrt{3}}{2}\Delta g_4 - \frac{\sqrt{3}}{2}\Delta g_5\right),$$

it follows that

$$\theta = \frac{1}{g}(-B_1^s + B_2^s)$$

$$= \frac{1}{\sqrt{3}g}(\Delta g_4 - \Delta g_2). \tag{8.10}$$

Of all the seven values of $\Delta g_0, \Delta g_1, \ldots \Delta g_6$, Δg_2 and Δg_4 only have effects on the value of θ at $x = D/2$. As an example, this method was applied to the average east–west profile of the northern part of Honshu, where

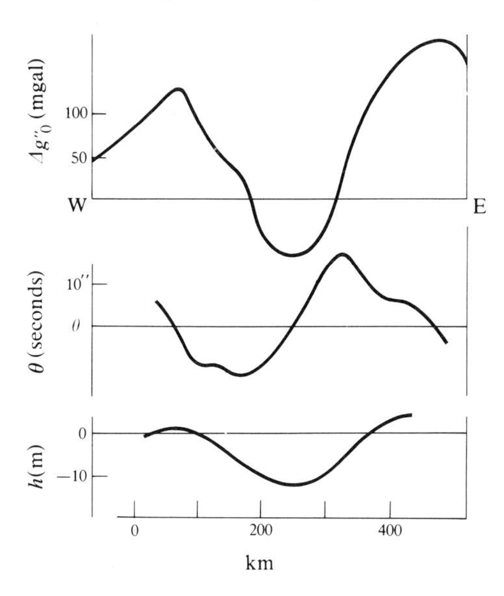

Figure 8.4 The Bouguer anomaly $\Delta g_0''$, deflection of the vertical θ, and undulation of the geoid h along the east–west direction across Honshu from the Sea of Japan coast to the Pacific Ocean coast.

Δg varies in the east–west direction only. A length of 585 km was taken and it was divided into nineteen points, 0, 1, 2, . . . 18. From Δg_0, Δg_1, . . . Δg_6, θ_3 was calculated, from Δg_1, Δg_2, . . . Δg_7, θ_4 was calculated, and so on by shifting one step each time. Thus the values of θ at twelve points, $x = 3, 4, \ldots 14$ are found. Integrating these with respect to x, the undulation amplitude h can be found. The results of the calculations are shown in Figure 8.4. The geoid is seen to have an undulation with a maximum amplitude of 10 m.

In the above example, 2π was divided by six only. The number of points can be increased if desired, but in that case the number of points at which the undulation of the geoid can be known decreases.

References and further reading

Tsuboi, C. 1937. The deflection of the vertical, the undulation of the geoid and the gravity anomalies. *Bull. Earthquake Res. Inst.* **15**, 650.

Tsuboi, C. 1940. Relation between the gravity anomalies and the corresponding subterranean mass distribution (V), Isostatic anomalies and the undulation of the isostatic geoid in the United States of America. *Bull. Earthquake Res. Inst.* **18**, 384.

Tsuboi, C. 1948. Undulation of the isostatic geoid in the East Indies as calculated from gravity anomalies. *Geophys. Notes*, no. 40.

Tsuboi, C. 1961. Numerical tables useful for studying the gravitational field at higher elevations of the Earth. In *The Earth today*, 73. Royal Astr. Soc.

Vos van Steenwijik, J. E. 1947. *Plumb-line deflections and geoid in Eastern Indonesia as derived from gravity*. Neth. Geod. Comm.

9 Gravity analysis in cylindrical co-ordinates

9.1 Advantage of using cylindrical co-ordinates

So far, various gravity problems have been discussed in cartesian co-ordinates with the aid of the Fourier series. The use of a Fourier series implies that a domain of 2π (one-dimensional case) or $2\pi \times 2\pi$ (two-dimensional case) is treated as a whole and every point in the domain is regarded as having equal weight in respective problems. There are many cases, however, in which we are interested only in one particular point in the domain. There are problems to find, for instance, the height of the geoid or the direction and magnitude of deflection of the vertical at a particular point from the gravity anomaly distribution observed around it. In these cases, it is more reasonable to refer to cylindrical co-ordinates, with the point of interest at its origin, than to refer to cartesian co-ordinates. In the two-dimensional Fourier series method, the domain taken in the analysis is a square with sides of $2a$. Measuring from the centre of the square, distances up to a are taken into consideration in the x- and y-directions, while those up to $\sqrt{2}a$ are taken in the diagonal directions. This uneven extension of the domain is not quite reasonable for dealing with gravity problems at the centre. In cylindrical co-ordinates, all the points equally distant from the centre are treated with equal weight. This is one of the fundamental reasons for the preference of cylindrical to cartesian co-ordinates.

9.2 Solution of Laplace's equation in cylindrical co-ordinates

Laplace's equation in cylindrical co-ordinates is

$$\frac{\partial^2 V}{\partial r^2} + \frac{1}{r}\frac{\partial V}{\partial r} + \frac{1}{r^2}\frac{\partial^2 V}{\partial \theta^2} + \frac{\partial^2 V}{\partial z^2} = 0,$$

as was shown before. For solving this equation, potential $V(r\theta z)$ will be assumed to be a product of $R(r)$, $\Theta(\theta)$, and $Z(z)$:

$$V(r\theta z) = R(r)\Theta(\theta)Z(z),$$

where $R(r)$ is a function of r only, $\Theta(\theta)$ a function of θ only and $Z(z)$ a function of z only, respectively. Putting this V in Laplace's equation, the following equation is obtained,

$$\frac{1}{R}\frac{\mathrm{d}^2R}{\mathrm{d}r^2} + \frac{1}{rR}\frac{\mathrm{d}R}{\mathrm{d}r} + \frac{1}{r^2\Theta}\frac{\mathrm{d}^2\Theta}{\mathrm{d}\theta^2} + \frac{1}{Z}\frac{\mathrm{d}^2Z}{\mathrm{d}z^2} = 0. \tag{9.1}$$

The last term on the left-hand side of this equation is a function of z only, while the other three terms are not. In spite of this, the sum of all the four terms must be zero. This is possible only when

$$\frac{1}{Z}\frac{\mathrm{d}^2Z}{\mathrm{d}z^2} = \pm k^2$$

and

$$\frac{1}{R}\frac{\mathrm{d}^2R}{\mathrm{d}r^2} + \frac{1}{rR}\frac{\mathrm{d}R}{\mathrm{d}r} + \frac{1}{r^2\Theta}\frac{\mathrm{d}^2\Theta}{\mathrm{d}\theta^2} = \mp k^2,$$

so that these two expressions are opposite in algebraic sign and cancel each other. Solutions to the first equation are generally

$$Z(z) = \exp(kz), \exp(-kz), \cos kz, \sin kz. \tag{9.2}$$

But in our case, in which z is taken vertically downwards with the origin $z = 0$ taken on the Earth's surface,

$$Z(z) = \exp(kz) \tag{9.3}$$

is the only meaningful solution near the Earth's surface. Then the remaining equation must be

$$\frac{1}{R}\frac{\mathrm{d}^2R}{\mathrm{d}r^2} + \frac{1}{rR}\frac{\mathrm{d}R}{\mathrm{d}r} + \frac{1}{r^2\Theta}\frac{\mathrm{d}^2\Theta}{\mathrm{d}\theta^2} = -k^2 \tag{9.4}$$

or

$$\frac{r^2}{R}\frac{\mathrm{d}^2R}{\mathrm{d}r^2} + \frac{r}{R}\frac{\mathrm{d}R}{\mathrm{d}r} + k^2r^2 = -\frac{1}{\Theta}\frac{\mathrm{d}^2\Theta}{\mathrm{d}\theta^2}.$$

The left-hand side of the last equation is a function of r only, while its right-hand side is solely a function of θ, and they must be equal. This is possible only when each side is equal to the same constants $\pm n^2$, so that

$$-\frac{1}{\Theta}\frac{d^2\Theta}{d\theta^2} = \pm n^2 \tag{9.5}$$

and

$$\frac{r^2}{R}\frac{d^2R}{dr^2} + \frac{r}{R}\frac{dR}{dr} + k^2r^2 = \pm n^2.$$

With the present co-ordinate axes, $+n^2$ only is appropriate, so that

$$\frac{d^2\Theta}{d\theta^2} + n^2\Theta = 0$$

and

$$\Theta(\theta) = \begin{matrix}\cos\\ \sin\end{matrix}\, n\theta.$$

The corresponding equation for $R(r)$ is then

$$\frac{d^2R}{dr^2} + \frac{1}{r}\frac{dR}{dr} + \left(k^2 - \frac{n^2}{r^2}\right)R = 0. \tag{9.6}$$

This last equation is called a Bessel's equation (F. W. Bessel, 1784–1846) and its solutions are written as $J_n(kr)$ with two parameters k and n. $J_n(kr)$ is called a **Bessel function** (of the first kind) of nth order. The constant k is a parameter which is related to the scale for measuring r. If kr is written as x, the Bessel's equation becomes

$$\frac{d^2J_n(x)}{dx^2} + \frac{1}{x}\frac{dJ_n}{dr} + \left(1 - \frac{n^2}{x^2}\right)J_n = 0. \tag{9.7}$$

Solutions $J_n(x)$ to this equation can be expressed by a power series of x:

$n = 0 \qquad J_0(x) = 1 - \frac{1}{4}x^2 + \frac{1}{64}x^4 - \ldots,$

$n = 1 \qquad J_1(x) = \frac{1}{2}x - \frac{1}{16}x^3 + \frac{1}{384}x^5 - \ldots,$

$n = 2 \qquad J_2(x) = \frac{1}{8}x^2 - \frac{1}{64}x^4 + \frac{1}{3072}x^6 - \ldots.$

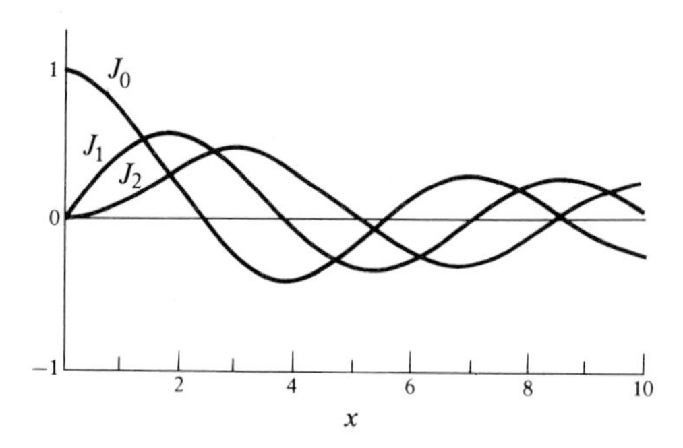

Figure 9.1 Curves of the Bessel functions, $J_0(x)$, $J_1(x)$ and $J_2(x)$.

The curves of $J_0(x)$, $J_1(x)$ and $J_2(x)$ are shown in Figure 9.1. Notice in particular that $x = 0$,

$$J_0(0) = 1 \qquad J_0'(0) = 0,$$

$$J_1(0) = 0 \qquad J_1'(0) = \tfrac{1}{2},$$

$$J_2(0) = 0 \qquad J_2'(0) = 0 \qquad J_2'(0) = \tfrac{1}{4}.$$

The contacts of the curves with the x-axis become higher in order according as n of J_n increases. As is seen in Figure 9.1 $J_0(x)$, $J_1(x)$ and $J_2(x)$ become zero at a number of points of x. $J_n(x) = 0$ has an infinite number of roots. Several of these roots are given in Table 9.1.

Table 9.1 Roots of $J_n(x) = 0$.

J_0	J_1	J_2
	0.000	0.000
2.405	3.832	5.135
5.520	7.016	8.417
8.654	10.173	11.620
11.792	...	...
...		

After all, the solution of Laplace's equation in cylindrical co-ordinates is a product of $J_n(kr)$, $\begin{matrix}\cos\\ \sin\end{matrix}\, n\theta$ and $\exp(kz)$, thus

$$V(r\theta z) = J_n(kr) \begin{matrix}\cos\\ \sin\end{matrix}\, n\theta \exp(kz). \tag{9.8}$$

But, since n can be any positive integer and k any positive number, the general expression of potential in cylindrical co-ordinates is

$$V(r\theta z) = \sum_n \int_0^\infty A_{nk} J_n(kr) \begin{matrix}\cos\\ \sin\end{matrix}\, n\theta \exp(kz)\, \mathrm{d}k,$$

where A_{nk} is a constant proper to each pair of n and k.

In this expression, $V(r\theta z)$ is separated into an r-dependent component, a θ-dependent component, and a z-dependent component. These three components, however, are not quite independent. The values of k in $J_n(kr)$ and in $\exp(kz)$ are the same and the values of n in $J_n(kr)$ and in $\begin{matrix}\cos\\ \sin\end{matrix}\, n\theta$ are the same. The first statement means that both r and z should be expressed in the same scale of length. The meaning of the second statement is as follows. Suppose a series of concentric circles are drawn on the plane $z = 0$ with the origin as the centre. Along each of the circles, the value of V varies with θ and can therefore be expressed by a Fourier series

$$V(r\theta) = \alpha_0(kr) + \alpha_1(kr)\cos\theta + \alpha_2(kr)\cos 2\theta + \ldots$$
$$+ \beta_1(kr)\sin\theta + \beta_2(kr)\sin 2\theta + \ldots. \qquad (9.9)$$

In this expression, the Fourier coefficients $\alpha_n(kr)$ and $\beta_n(kr)$ change according to the r of the circles and their modes of variation are the same as those of $J_n(kr)$. For instance, $\alpha_0(kr)$ changes with r in the form as $J_0(kr)$ does, $\alpha_1(kr)$ and $\beta_1(kr)$ as $J_1(kr)$ does, and $\alpha_2(kr)$ and $\beta_2(kr)$ as $J_2(kr)$ does, and so on. The product of $\begin{matrix}\cos\\ \sin\end{matrix}\, n\theta \cdot J_n(kr)$ becomes zero along several nodal lines and nodal circles as shown in Figure 9.2. The nodal lines are where $\begin{matrix}\cos\\ \sin\end{matrix}\, n\theta = 0$ and the nodal circles are where $J_n(kr) = 0$.

With an appropriate constant A_{nk} multiplied by $\begin{matrix}\cos\\ \sin\end{matrix}\, n\theta \cdot J_n(kr)$, the sum

$$\sum_n A_{nk} \begin{matrix}\cos\\ \sin\end{matrix}\, n\theta J_n(kr)$$

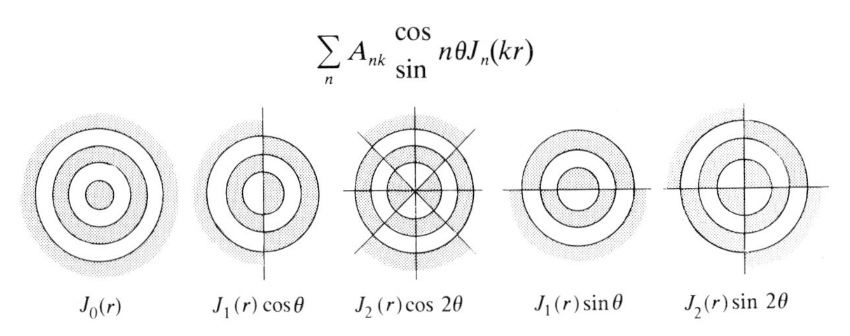

Figure 9.2 Nodal circles and nodal lines of $J_n(r) \begin{matrix}\cos\\ \sin\end{matrix}\, n\theta$.

can represent a distribution of any continuous quantity given on the plane $z = 0$. This expression is called a **Fourier–Bessel series**. Since k is not fixed to be a single value, the above expression becomes generally

$$\sum_n \int_0^\infty A_{nk} \frac{\cos}{\sin} n\theta J_n(kr)\, \mathrm{d}r.$$

9.3 Height of the geoid in cylindrical co-ordinates

Now that the expression for the distribution of potential V has been obtained, let us use it for studying several actual gravity problems. In this section, a method will be shown to find the height of geoid $h(0)$ from a distribution of Δg which is known around the point $r = 0$. The height of the geoid is its distance from the undisturbed geoid in case there is no anomaly Δg.

If the potential at the origin is $V(0)$, then the height of the geoid is given by

$$h(0) = V(0)/g. \tag{9.10}$$

So our problem is how to express $V(0)$ in terms of $\Delta g(r\theta)$ which is known. Generally $V(r\theta z)$ is

$$V(r\theta z) = \sum_n \int_0^\infty A_{nk} J_n(kr) \frac{\cos}{\sin} n\theta \exp(kz)\, \mathrm{d}k. \tag{9.11}$$

Since at $r = 0$, the potential does not have a component which varies with θ, all $J_n(kr)$ with the value of n other than $n = 0$ do not come into the problem at the origin. Also on the Earth's surface, z is zero, so that $\exp(kz)$ is one. The expression for $V(0)$ will become then

$$V(0) = \int_0^\infty A_{0k} J_0(kr)\, \mathrm{d}k.$$

But, since $J_0(0) = 1$, this becomes finally

$$V(0) = \int_0^\infty A_{0k}\, \mathrm{d}k. \tag{9.12}$$

By this expression, $V(0)$ can be found if A_{0k} is calculated from the observed distribution of Δg. Now, Δg is the vertical gradient of V and is given by

$$\Delta g = \frac{\partial v}{\partial z} = \sum_n \int_0^\infty A_{nk} k J_n(kr) \begin{matrix}\cos\\ \sin\end{matrix} n\theta \exp(kz)\, \mathrm{d}k.$$

Putting $n = 0$, the mean value $\overline{\Delta g(r)}$ on a circle $r = r$ is

$$\overline{\Delta g(0)} = \int_0^\infty k A_{0k} J_0(kr)\, \mathrm{d}k.$$

On the other hand, the observed variation of $\overline{\Delta g(r)}$ according to r must also be expressed as

$$\overline{\Delta g(r)} = \int_0^\infty B_{0k} J_0(kr)\, \mathrm{d}k.$$

By equating $\overline{\Delta g(r)}$ in the above two expressions, it is seen that

$$A_{0k} = \frac{B_{0k}}{k}. \tag{9.13}$$

By this relationship, A_{0k} in the expression for the potential $V(0)$ can be found from B_{0k} in the expression for the gravity distribution. There is a mathematical relation

$$\int_0^\infty J_0(kr)\, \mathrm{d}r = \frac{1}{k}$$

so that

$$V(0) = \int_0^\infty A_{0k}\, \mathrm{d}k$$

$$= \int_0^\infty \frac{B_{0k}}{k}\, \mathrm{d}k,$$

$$V_{(0)} = \int_0^\infty \mathrm{d}r \int_0^\infty B_{0k} J_0(kr)\, \mathrm{d}k$$

$$= \int_0^\infty \overline{\Delta g(r)}\, \mathrm{d}r.$$

The height of a geoid at $r = 0$ is therefore

$$h(0) = \frac{V(0)}{g} = \frac{1}{g}\int_0^\infty \overline{\Delta g(r)}\,dr. \tag{9.14}$$

This is a very simple relation to handle. To summarize:

(a) draw a series of concentric circles with the point in question as the centre;

(b) find $\overline{\Delta g(r)}$ which is the mean value of Δg on each circle;

(c) integrate $\overline{\Delta g(r)}$ with respect to r;
(d) divide finally the integrated value by g (which equals 980).

The answer will give the height of the geoid at the origin $r = r$. Along the circles with larger r, Δg will be positive as well as negative, so that $\overline{\Delta g(r)}$ will tend to zero and $\int_0^\infty \overline{\Delta g(r)}\,dr$ will converge to a certain asymptotic value.

As a hypothetical simple example, let us assume that $\overline{\Delta g(r)}$ is expressed by

$$\overline{\Delta g(r)} = \frac{\Delta g(0)}{2}\left(1 + \cos \pi \frac{r}{a}\right),$$

where $0 \leqslant r \leqq a$

and $$\overline{\Delta g(r)} = 0,$$

where $a \leqslant r$, as shown in Figure 9.3. The height of the geoid at $r = 0$ is then

$$h(0) = \frac{\Delta g(0)}{2g}\int_0^a \left(1 + \cos \pi \frac{r}{a}\right) dr$$

$$= \frac{\Delta g(0)}{2g} a. \tag{9.15}$$

The rise of the geoid is proportional to $\Delta g(0)$ and, for a given $\Delta g(0)$, it is proportional to a. Table 9.2 gives the values of $h(0)$ for various values of $\Delta g(0)$ and a.

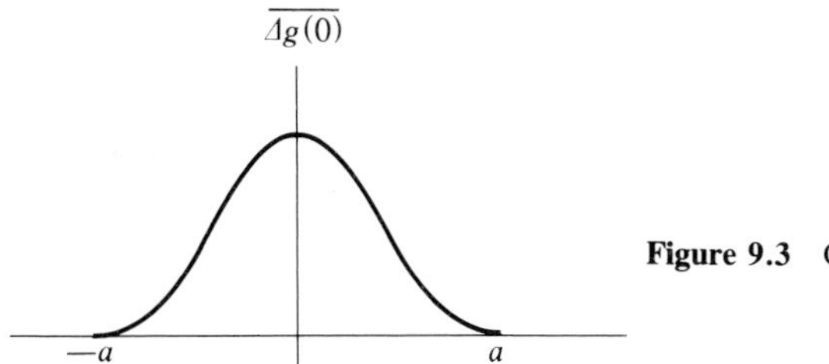

Figure 9.3 Curve of $\overline{\Delta g(r)} = \dfrac{\Delta g(0)}{2}\left(1 + \cos 2\pi \dfrac{r}{a}\right)$.

Table 9.2 The height of the geoid for various $\Delta g(0)$ and a.

$\Delta g(0) \backslash a$	1 km	10 km	100 km
1 mgal	0.05 cm	0.5 cm	5.1 cm
10 mgal	0.5 cm	5.1 cm	51 cm
100 mgal	5.1 cm	51 cm	510 cm

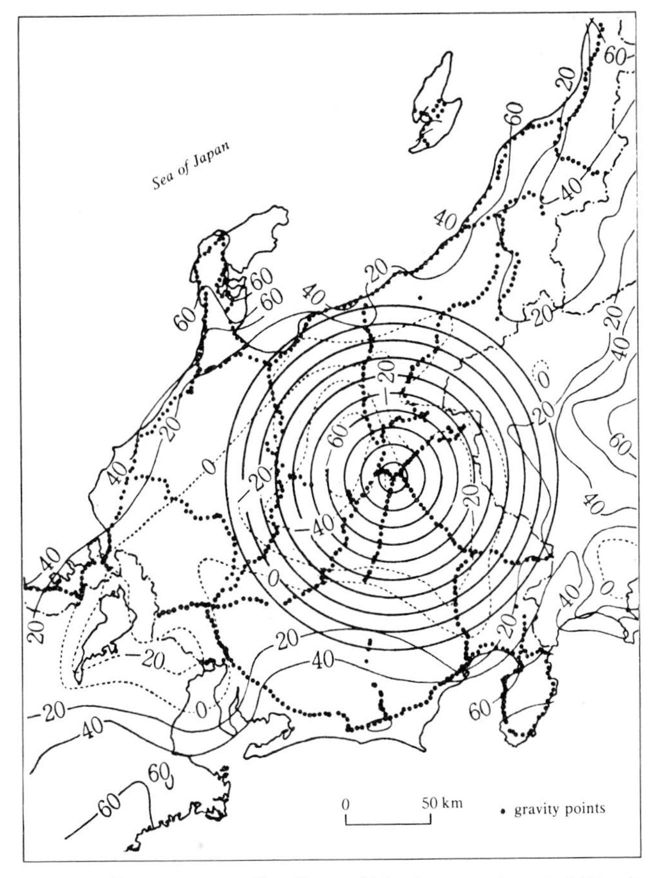

Figure 9.4 Bouguer anomalies (in mgal) in the central part of Honshu.

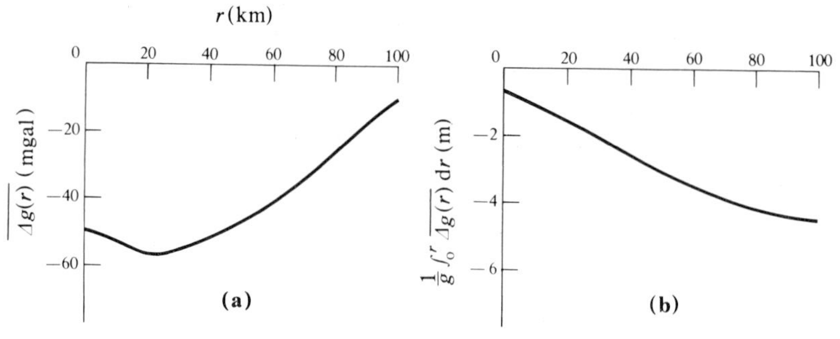

Figure 9.5

The following is an actual example of the application of the above method. The distribution of Bouguer anomalies $\Delta g_0''$ in the central part of Honshu, Japan is as shown in Figure 9.4. Taking a point $\phi = 36°$N, $\lambda = 138°$E as the centre, the mean values of $\Delta g_0''$ on the circles $r = 10$, 20, . . . 100 km are plotted in Figure 9.5a. The curve of Figure 9.5b is the result of the integration of $\overline{\Delta g_{(r)}}$ with respect to r divided by $2g$. This curve tends to -4.8 m, showing that the Bouguer geoid at the origin is 4.8 m lower, compared with those at large distances.

9.4 Deflection of the vertical in cylindrical co-ordinates

In this section, a method will be developed for finding the deflection of the vertical at a point from the variation of Δg around it (Tsuboi 1954b). The potential V in cylindrical co-ordinates is

$$V(r\theta z) = \sum_n \int_0^\infty A_{nk}{}^c J_n(kr) \cos n\theta \exp(kz)\, dk$$

$$+ \sum_n \int_0^\infty A_{nk}{}^s J_n(kr) \sin n\theta \exp(kz)\, dz \qquad (9.16)$$

where the z-axis is taken into the undisturbed downward direction. Δg on the Earth's surface $z = 0$ is given by:

$$\Delta g(r\theta 0) = \left| \frac{\partial V}{\partial z} \right|_{z=0}$$

$$= \int_0^\infty k A_{0k}{}^c J_0(kr)\, dk$$

$$+\int_0^\infty kA_{1k}{}^cJ_1(kr)\,dk\cdot\cos\theta+\int_0^\infty kA_{2k}{}^cJ_2(kr)\,dk\cdot\cos 2\theta+\ldots$$

$$+\int_0^\infty kA_{1k}{}^sJ_1(kr)\,dk\cdot\sin\theta+\int_0^\infty kA_{2k}{}^sJ_2(kr)\,dk\cdot\sin 2\theta+\ldots \tag{9.17}$$

and this can be written as

$$\Delta g(r\theta 0)=C_0+C_1(r)\cos\theta+C_2(r)\cos 2\theta+\ldots$$
$$+S_1(r)\sin\theta+S_2(r)\sin 2\theta+\ldots \tag{9.18}$$

where

$$C_1(r)=\int_0^\infty A_{1k}{}^cJ_1(kr)\,dk,$$

$$S_1(r)=\int_0^\infty A_{1k}{}^sJ_1(kr)\,dk.$$

$C_n(r)$ and $S_n(r)$ are the coefficients of $\cos n\theta$ and $\sin n\theta$ when the distribution of Δg along a circle of radius r is expressed in a Fourier series with respect to θ. The eastward component ξ and the northward component η of the deflection of the vertical at the origin are

$$\xi=\frac{1}{g}\left(\frac{\partial V}{\partial r}\right)_{\theta=0}$$

$$=\frac{1}{g}\left\{\int_0^\infty kA_{0k}{}^cJ_0'(0)\,dk+\int_0^\infty kA_{1k}{}^cJ_1'(0)\,dk+\int_0^\infty kA_{2k}{}^cJ_2'(0)\,dk+\ldots\right\},$$

$$\eta=\frac{1}{g}\left(\frac{\partial V}{\partial r}\right)_{\theta=\pi/2}$$

$$=\frac{1}{g}\left\{\int_0^\infty kA_{1k}{}^sJ_1'(0)\,dk+\int_0^\infty kA_{2k}{}^sJ_2'(0)\,dk+\ldots\right\},$$

respectively. But since

$$J_0'(0)=0\qquad J_1'(0)=\tfrac{1}{2}\qquad J_2'(0)=J_3'(0)=\ldots 0$$

as stated before, ξ and η will become simply

$$\xi = \frac{1}{2g}\int_0^\infty kA_{1k}^{c}\,\mathrm{d}k$$

$$\eta = \frac{1}{2g}\int_0^\infty kA_{1k}^{s}\,\mathrm{d}k.$$

Using a property of the Bessel function that

$$\int_0^\infty \frac{J_1(kr)}{r}\,\mathrm{d}r = 1,$$

we see that

$$\begin{aligned}\int_0^\infty kA_{1k}^{c}\,\mathrm{d}k &= \int_0^\infty\!\!\int_0^\infty kA_{1k}^{c}\frac{J_1(kr)}{r}\,\mathrm{d}k\,\mathrm{d}r\\ &= \int_0^\infty \frac{C_1(r)}{r}\,\mathrm{d}r \qquad (9.20)\\ \int_0^\infty kA_{1k}^{s}\,\mathrm{d}k &= \int_0^\infty\!\!\int_0^\infty kA_{1k}^{s}\frac{J_1(kr)}{r}\,\mathrm{d}k\,\mathrm{d}r\\ &= \int_0^\infty \frac{S_1(r)}{r}\,\mathrm{d}r\end{aligned}$$

and

$$\xi = \frac{1}{2g}\int_0^\infty \frac{C_1(r)}{r}\,\mathrm{d}r,$$

$$\eta = \frac{1}{2g}\int_0^\infty \frac{S_1(r)}{r}\,\mathrm{d}r.$$

$C_1(r)$ and $S_1(r)$ in these expressions are the coefficients of the first-order cosine terms and sine terms respectively when Δg values along the circle $r = r$ around the origin are expressed in a Fourier series of azimuth θ. To summarise, the procedure by which the values of ξ and η are found is as follows:

(a) draw a series of concentric circles around the origin;
(b) calculate the first-order coefficients $C_1(r)$ and $S_1(r)$, when $\Delta g(r)$ is expressed in a Fourier series of θ;

(c) divide $C_1(r)$ and $S_1(r)$ by r;
(d) integrate $C_1(r)/r$ and $S_1(r)/r$ from $r = 0$ according to r until they tend to asymptotic values;
(e) divide these asymptotic values by $2g$.

These will give the values of ξ and η at the origin. Although r comes in the denominators of $C_1(r)/r$ and $S_1(r)/r$, they will not diverge at $r = 0$. The reason for this is that $C_1(r)$ and $S_1(r)$ themselves change as

$$J_1(r) = \frac{1}{2}\left(r - \frac{r^3}{8} + \ldots\right) \tag{9.21}$$

and are zero at $r = 0$. For very small r, $C_1(r)$ and $S_1(r)$ change as the first power of r and the integration of $C_1(r)/r$ and $S_1(r)/r$ can be divided into two parts, from $r = 0$ to $r = R$ and from $r = R$ to $r = a$. In the first part, $C_1(r)$ and $S_1(r)$ change as αr so that

$$\begin{aligned}\xi &= \frac{1}{2g}\int_0^a \frac{C_1(r)}{r}\,\mathrm{d}r \\ &= \frac{1}{2g}\left\{\int_0^R \frac{C_1(r)}{r}\,\mathrm{d}r + \int_R^a \frac{C_1(r)}{r}\,\mathrm{d}r\right\} \\ &= \frac{1}{2g}\left\{\alpha R + \int_R^a \frac{C_1(r)}{r}\,\mathrm{d}r\right\} \\ &= \frac{1}{2g}\left\{C_1(R) + \int_R^a \frac{C_1(r)}{r}\,\mathrm{d}r\right\}.\end{aligned}$$

Similarly,

$$\eta = \frac{1}{2g}\left\{S_1(R) + \int_R^a \frac{S_1(r)}{r}\,\mathrm{d}r\right\}.$$

As an example of an application of the above mentioned, the deflection of the vertical at Tokyo, Japan, will be discussed. It has been known that the position of Tokyo (the old site of the Tokyo Astronomical Observatory) as determined by astronomical measurements is not exactly compatible with that derived from recent observations of the occultations of stars or of the positions of artificial satellites. The latitude of Tokyo derived from astronomical data is no less than 10″–12″ too small in angle and its eastern longitude is too large by about the same amount. These discrepancies are produced by

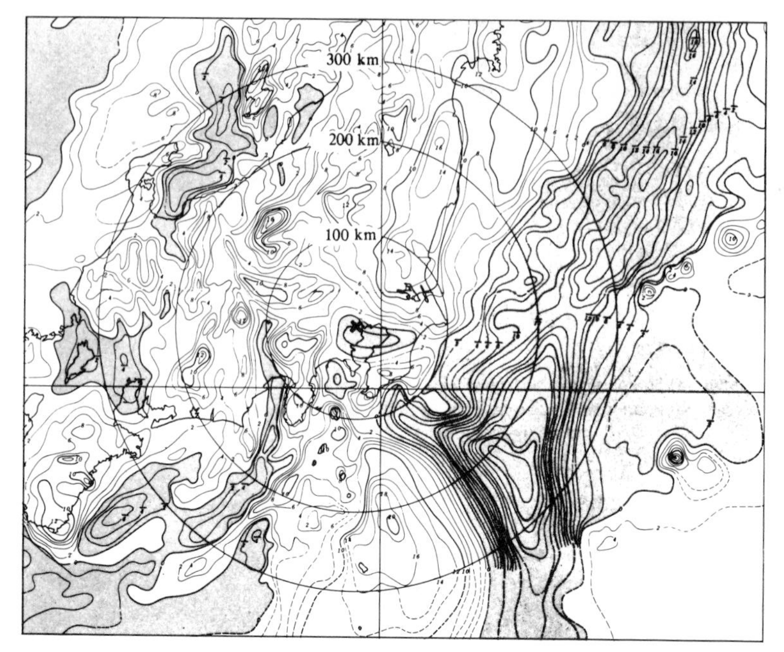

Figure 9.6 Distribution of Δg_0 around Tokyo.

the deflection of the vertical. In the classical astronomical methods for determining the position of a point on the Earth's surface, the direction of the vertical at that point is the essential reference which is indicated by a spirit level attached to a telescope to observe the stars. In the modern methods for observing artificial satellites or occultations of stars, such reference is not needed. Let us calculate the deflection of the vertical at Tokyo from the gravity anomalies observed around Tokyo and see if the deflection so calculated agrees with the suspected value.

The distribution of Δg_0 around the city is shown in Figure 9.6. A series of concentric circles were drawn at 20 km intervals, with Tokyo at the

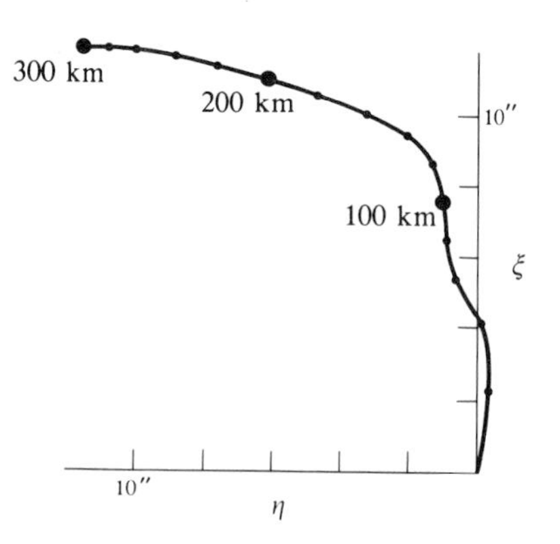

Figure 9.7 ξ- and η-values vary according to r.

centre, and calculations as described above were made. Figure 9.7 shows how ξ- and η-components vary as the limit of integration is increased. The ξ- and η-components appear to tend closer to $10''$–$12''$ as expected.

It should be noted that the ξ-component increases rather rapidly up to 120 km, while the η-component does not do so within that distance, but then increases rapidly beyond that distance. The former variation is caused by the mass deficiency to the immediate south of Tokyo, due to the basin structure of young and thick sediments. The latter variation is due to the effect of the big mass deficiency structure off the Pacific coast of Honshu, Japan.

References and further reading

Tsuboi, C. 1954a. Calculation of the deflection of the vertical with the aid of Bessel–Fourier series (in Japanese). *J. Geod. Soc., Japan* **1**, 25.

Tsuboi, C. 1954b. A simple method for calculating the deflection of the vertical from gravity anomalies, with its applications to 16 selected stations in the USA. *J. Phys. Earth* **2**, 45.

Tsuboi, C. 1954c. A new and simple method for calculating the deflections of the vertical from gravity anomalies with the aid of the Bessel–Fourier series. *Proc. Jap. Acad.* **30**, 461.

Tsuboi, C. 1961a. Upward and downward continuation of gravity values based on the cylindrical co-ordinate system. *Proc. Jap. Acad.* **37**, 37.

Tsuboi, C. 1961b. *Upward continuation of gravity values based on the cylindrical co-ordinate system*. Rep. Inst. Geod. Photogr. Cartogr. Ohio State University, no. 16.

Tsuboi, C. 1979. Deflection of the vertical at Tokyo. *Proc. Jap. Acad.* **55**, 49.

10 Undulation of the geoid and deflection of the vertical in spherical co-ordinates

10.1 Undulation of the geoid

Theoretical relations existing between the gravity anomaly distribution on a spherical earth and the corresponding undulation of the geoid were discussed by G. C. Stokes (1819–1903) more than a hundred years ago. He showed that if a gravity anomaly distribution on the Earth's surface is expressed in a spherical harmonic series such as

$$\Delta g = g(u_2 + u_3 + u_4 + \ldots) \tag{10.1}$$

then, the height of the geoid is given by

$$h = R\left(u_2 + \frac{u_3}{2} + \frac{u_4}{3} + \ldots\right)$$

where R is the mean radius of the Earth. This is a very fine mathematical relationship, but in order that this relation can be used in actual problems, a full knowledge of gravity distribution over the whole Earth is needed or an appropriate process of interpolation of gravity values is necessary for filling in those areas where few gravity data are available.

Pizetti (1911) and Lambert (1931) transformed Stokes's expression into a more practicable form by using zonal harmonic series instead of spherical harmonic series. The point O at which the height of the geoid is to be calculated is taken as the pole of zonal co-ordinate as shown in

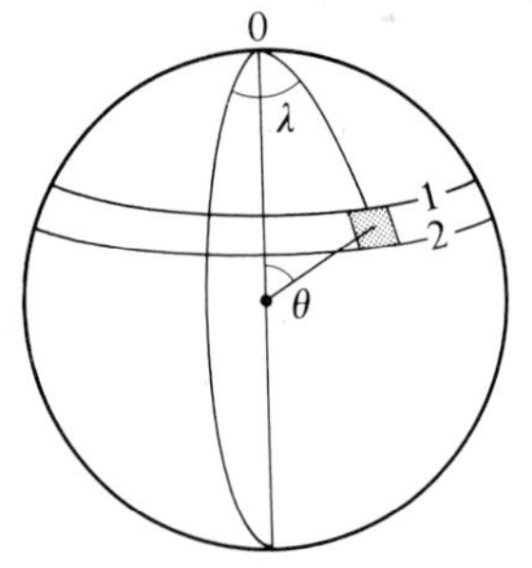

Figure 10.1 Spherical co-ordinates.

Figure. 10.1. The surface of the Earth is divided into a number of compartments $R\Delta\theta \times R \sin \theta\Delta\lambda$, where θ is the co-latitude and λ the azimuth of the compartment as seen from the pole. The effect of Δg in a compartment on the height of the geoid at O is a function of θ only. If the azimuthal average value of Δg taken along a zonal belt between θ and $(\theta + \Delta\theta)$ is expressed by $\overline{\Delta g(\theta)}$, then the height of the geoid at O is shown to be given by

$$h = \frac{R}{g}\int_0^\pi \overline{\Delta g(0)}\, f(\theta)\sin\theta\, d\theta, \qquad (10.2)$$

the integration being extended to the whole Earth. $f(\theta)$ has a complicated form such as

$$f(\theta) = \frac{1}{2}\left[\operatorname{cosec}\frac{\theta}{2} + 1 - 6\sin\frac{\theta}{2} - 5\cos\theta - 3\cos\theta\log_e\left\{\sin\frac{\theta}{2}\left(1 + \sin\frac{\theta}{2}\right)\right\}\right].$$

If a summation is used instead of integration in the above expression, it becomes

$$h = \frac{R}{g}\Sigma\,\overline{\Delta g(\theta)}\,\left[\overline{f(\theta)\sin\theta}\right]_{\theta 1}^{\theta 2}(\theta_1 - \theta_2),$$

where

$$\left[\overline{f(\theta)\sin\theta}\right]_{\theta 1}^{\theta 2}$$

is the average value of $f(\theta)\sin\theta$ within a belt between $\theta = \theta_1$ and $\theta = \theta_2$. If we put

$$\int f(\theta)\sin\theta\, d\theta = \Phi(\theta)$$

h is given by

$$h = \frac{R}{g}\Sigma\,\overline{\Delta g}\,[\Phi(\theta_2) - \Phi(\theta_1)],$$

where

$$\Phi(\theta) = \frac{1}{2}\left[4\sin\frac{\theta}{2} + 2\sin^2\frac{\theta}{2} - 6\sin^3\frac{\theta}{2} - \frac{7}{4}\sin^2\theta - \frac{3}{2}\sin^2\theta\log_e\left\{\sin\frac{\theta}{2}\left(1 + \sin\frac{\theta}{2}\right)\right\}\right].$$

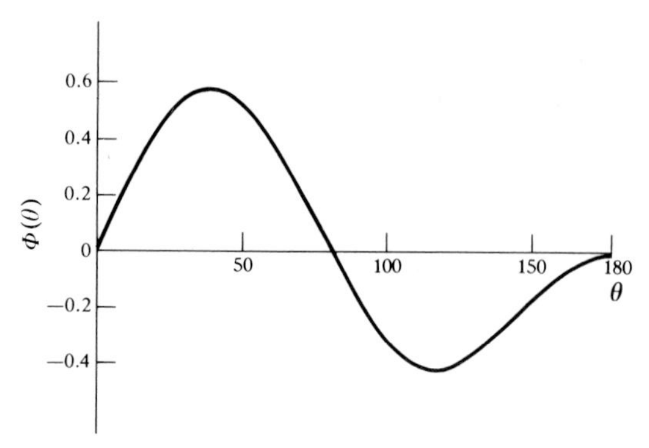

Figure 10.2 Values of $\phi(\theta)$ according to θ.

The curve in Figure 10.2 shows the values of $\Phi(\theta)$ according to θ. Table 10.1 gives the value of $\Phi(\theta)$ and $\{\Phi(\theta + 10°) - \Phi(\theta)\}$ for various values of θ and also the value of h at the point O in a case where the average of Δg within a belt of 10° is 1 mgal.

Table 10.1 Values of $\Phi(\theta)$ and $\{\Phi(\theta + 10°) - \Phi(\theta)\}$ for various values of θ.

Co-latitude θ	$\Phi(\theta)$	$\Phi(\theta + 10°) - \Phi(\theta)$	h (cm)
0°	0.0000	0.2068	134
10°	+0.2068	0.1921	125
20°	+0.3989	0.1252	82
30°	+0.5241	0.0367	24
40°	+0.5608	−0.0530	−35
50°	+0.5078	−0.1272	−83
60°	+0.3806	−0.1753	−114
70°	+0.2053	−0.1915	−125
80°	+0.0138	−0.1764	−115
90°	−0.1626	−0.1355	−88
100°	−0.2981	−0.0733	−48
110°	−0.3714	−0.0214	−14
120°	−0.3928	0.0394	26
130°	−0.3534	0.0789	52
140°	−0.2745	0.0967	63
150°	−0.1778	0.0908	59
160°	−0.0870	0.0640	42
170°	−0.0230	0.0230	19

For example, if the average value of Δg along a belt between $\theta = 30°$ and $\theta = 40°$ is 10 mgal, the effect of this belt on the height of the geoid at 0 is

$$10 \times 24 = 240 \text{ cm.}$$

In this way, the height of the geoid at the point O can be calculated from the average values of Δg taken along all the zones surrounding O. But to do so, a full knowledge of the distribution of Δg over the whole Earth is needed and when the point O is changed, the corresponding zonal averages of Δg should be recalculated each time. This is an elaborate thing to do.

Since the distribution of Δg on the Earth's surface is irregular $\overline{\Delta g\theta}$ will generally become smaller and smaller as θ is increased. In an approximate calculation, therefore, it will cause no serious harm if the integration with respect to θ is stopped at a certain limiting value, instead of extending θ to π. If θ is small,

$$f(\theta)\sin\theta \approx 1$$

and since

$$r = R\theta \qquad \mathrm{d}r = R\,\mathrm{d}\theta,$$

we see

$$\begin{aligned} h &= \frac{R}{g}\int_0^{\theta} \overline{\Delta g(\theta)}\, f(\theta)\sin\theta\,\mathrm{d}\theta \\ &= \frac{R}{g}\int_0^{r} \overline{\Delta g(r)}\,\frac{\mathrm{d}r}{R} \\ &= \frac{1}{g}\int_0^{r} \overline{\Delta g(r)}\,\mathrm{d}r \end{aligned} \tag{10.3}$$

and this expression is the same as the one derived before by using the Bessel function.

10.2 Deflection of the vertical in spherical co-ordinates

With regard to the deflection of the vertical, Vening Meinesz (1928) discovered a method for calculating it from the distribution of Δg over the whole surface of the Earth. According to him, the east–west and north–south components of deflection of the vertical at a point O are given by

$$\xi = \frac{1}{2\pi g}\int_0^{2\pi}\mathrm{d}\lambda\int_0^{\pi}\Delta g(\theta\lambda)\,\frac{\partial f(\theta)}{\partial\theta}\sin\theta\cos\lambda\,\mathrm{d}\theta, \tag{10.4}$$

$$\eta = \frac{1}{2\pi g}\int_0^{2\pi} \mathrm{d}\lambda \int_0^{\pi} \Delta g(\theta\lambda)\, \frac{\partial f(\theta)}{\partial \theta} \sin\theta \sin\lambda \, \mathrm{d}\theta, \tag{10.5}$$

where $\partial f(\theta)/\partial\theta$ has a complicated form such as

$$\frac{\partial f(\theta)}{\partial \theta}\sin\theta = \frac{1}{2}\left\{-\operatorname{cosec}\frac{\theta}{2} - 3 - 8\sin\frac{\theta}{2} + 32\sin^2\frac{\theta}{2}\right.$$

$$\left. + 12\sin^3\frac{\theta}{2} - 32\sin^4\frac{\theta}{2} + 3\sin^2\theta \log_e\left(\sin\frac{\theta}{2} + \sin^2\frac{\theta}{2}\right)\right\}.$$

The values of $\{\partial f(\theta)/\partial\theta\}\sin\theta$ were given by Sollins (1947), in the form of a numerical table. Here again, a full knowledge of the gravity anomaly distribution over the whole surface of the Earth is needed.

10.3 Rice's calculations and their simplification

Donald A. Rice (1949) of the US Coast and Geodetic Survey elaborately applied the above method to find deflections of the vertical at 16 geodetic points in the USA. Being unable to extend the integration over the whole surface of the Earth due to lack of data, he did not extend the integration to π. The results of his calculations for the geodetic points at Dirks are shown in Figure 10.3 by the marks ⊙. Seeing that both ξ and η values of the deflection of the vertical become nearly constant, he stopped the integration at a distance of 400 km.

At a distance of 400 km, the separation of the actual Earth's surface and a tangential geometrical plane drawn at the point O is only 13 km,

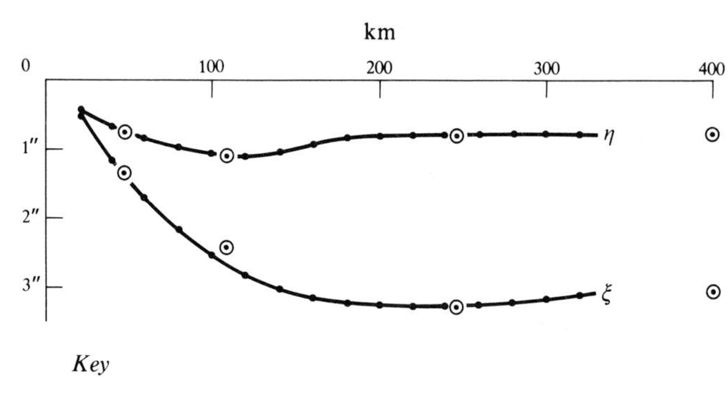

Figure 10.3 Deflection of the vertical at Dirks.

so that an approximation is allowed to neglect the curvature of the Earth and treat the problem as a plane problem by using the Fourier–Bessel series which was explained in the preceding chapter. The points marked ● in Figure 10.3 were calculated by that method and the same gravity data as Rice were used. The points were taken at 20 km steps from 0 to 320 km. The points ⦿ and ● in Figure 10.3 agree well (Tsuboi 1954).

10.4 Comparison of the spherical co-ordinate method and the cylindrical co-ordinate method

As was shown in the preceding section, the results obtained using the two different co-ordinates agree well. This is easily deduced, as will be shown below.

If θ is small

$$\frac{\partial f(\theta)}{\partial \theta} = -\frac{1}{\theta^2},$$

$$\frac{\partial f(\theta)}{\partial \theta} \sin\theta = \frac{1}{\theta}, \tag{10.6}$$

so that Vening Meinesz's formula will become

$$\xi = \frac{1}{2\pi g}\int_0^{2\pi} \mathrm{d}\lambda \int_0^{\theta} \Delta g(\theta\lambda)\,\frac{1}{\theta}\cos\lambda\,\mathrm{d}\theta, \tag{10.7}$$

$$\eta = \frac{1}{2\pi g}\int_0^{2\pi} \mathrm{d}\lambda \int_0^{\theta} \Delta g(\theta\lambda)\,\frac{1}{\theta}\sin\lambda\,\mathrm{d}\theta. \tag{10.8}$$

In the Fourier–Bessel series expressions,

$$\frac{1}{\pi}\int_0^{2\pi} \Delta g(\theta\lambda)\cos\lambda\,\mathrm{d}\lambda = C_1(\theta),$$

$$\frac{1}{\pi}\int_0^{2\pi} \Delta g(\theta\lambda)\sin\lambda\,\mathrm{d}\lambda = S_1(\theta).$$

Putting these in the expressions for ξ and η, we obtain

$$\xi = \frac{1}{2g}\int_0^{\theta} \frac{C_1(\theta)}{\theta}\,\mathrm{d}\theta,$$

$$\eta = \frac{1}{2g}\int_0^\theta \frac{S_1(\theta)}{\theta}\,d\theta.$$

If the integration with respect to θ is stopped at $\theta = \theta$ instead of π, then

$$\xi = \frac{1}{2g}\int_0^\theta \frac{C_1(\theta)}{\theta}\,d\theta,$$

$$\eta = \frac{1}{2g}\int_0^\theta \frac{S_1(\theta)}{\theta}\,d\theta,$$

but $r = R\theta$ where R is the radius of the Earth, then

$$\xi = \frac{1}{2g}\int_0^r \frac{C_1(r)}{r}\,dr,$$

$$\eta = \frac{1}{2g}\int_0^r \frac{S_1(r)}{r}\,dr,$$

which are simply the expressions for cylindrical co-ordinates in the preceding chapter (Tsuboi & Hayatsu 1954).

10.5 Artificial satellites and the shape of the Earth

The introduction of artificial satellites has opened a new way for finding the shape of the geoid, by investigating their orbits. The gravity potential of the Earth expressed in a zonal harmonic series is

$$V = \frac{GM}{r}\left\{1 - \frac{J_2}{r^2}P_2(\cos\theta) - \frac{J_3}{r^3}P_3(\cos\theta) + \ldots\right\}. \tag{10.9}$$

According to Y. Kozai in 1969, the values of J_2 and J_3 are

$$J_2 = 1.08264 \times 10^{-3}$$

$$J_3 = -2.54 \times 10^{-6}.$$

The results show that the coefficient J_3 of the term $P_3(\cos\theta)$, which is asymmetric with respect to the equatorial plane, is not negligibly small. As a result of this, the shape of the geoid is not symmetrical with respect to the equatorial plane and is like a pear, as shown in Figure 10.4. At the North Pole, the geoid is 13.5 m higher than an ellipsoid and at the

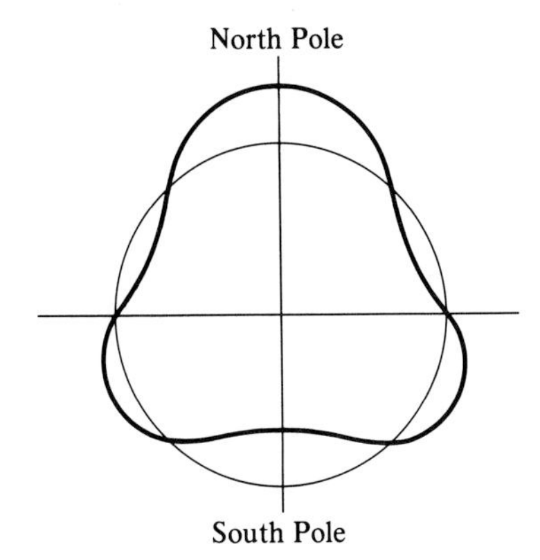

Figure 10.4 The pear-shaped geoid.

South Pole, it is 24.1 m lower than that. Many investigations are now being made to determine the shape of the geoid accurately from satellite observations.

References and further reading

Kozai, Y. 1969. Revised values for coefficients of zonal spherical harmonics in the geopotential. In *Dynamics of satellites*, 104. Berlin: Springer.

Lambert, W. D. 1947. Deflections of the vertical from gravity anomalies. *Trans. Am. Geophys. Union* **28**, 153.

Lambert, W. D. and F. W. Darling 1936. *Tables for determining the form of the geoid and its indirect effect on gravity*. Spec. Pub. USCGS, no. 199.

Pizetti, P. 1911. Sopra il calcolo delle deviazione del geoide dell'ellissoide. *Atti Reale Accad. Sci., Torino* **46**, 223.

Rice, D. A. 1949. Gravimetric deflections by the method of condensation. *Trans Am. Geophys. Union* **30**, 323.

Rice, D. A. 1952. Deflection of the vertical from gravity anomalies. *Bull. Geod.* **25**, 285.

Sollins, A. D. 1947. Tables for the computation of deflections of the vertical from gravity anomalies. *Bull. Geod.* **6**, 279.

Stokes, G. C. 1849. On the variation of gravity at the surface of the Earth. *Trans. Camb. Phil Soc.* **8**, 672.

Vening Meinesz, F. A. 1928. A formula expressing the deflection of the plumb-line in the gravity anomalies and some formulae for the gravity field and gravity potential outside the geoid. *Kon. Akad. van Wet., Amsterdam* **31**, 315.

11 Second derivatives of the gravity potential

11.1 Meanings of the second derivatives of gravity potential

The gravity force is the first derivative of the gravity potential V in the vertical direction, $\partial V/\partial z$. Since the z-axis is taken vertically downwards at the origin, both $\partial V/\partial x$ and $\partial V/\partial y$ in horizontal directions are zero. There is thus only one first derivative of potential that is not zero.

There are six second derivatives of gravity potential, namely:

(a) $\partial^2 V/\partial x^2$, $\partial^2 V/\partial x\, \partial y$, $\partial^2 V/\partial y^2$,
(b) $\partial^2 V/\partial z\, \partial x$, $\partial^2 V/\partial y\, \partial z$,
(c) $\partial^2 V/\partial z^2$.

As will be seen later, the three terms in (a) are related to the curvature of the geoid and are called curvature terms. The two terms in (b) are the rates of variation of g in horizontal directions and are called gradient terms. The term in (c) is the rate of variation of g in the vertical direction.

Let us see, in the first place, what relationship generally exists between the first and second derivatives by comparing $\partial V/\partial z$ and $\partial^2 V/\partial z\, \partial x$ as an example. Suppose a distribution of g on the Earth's surface is expressed by

$$g(x) = \frac{\partial V}{\partial z} = \sum_m B_m \begin{matrix}\cos\\ \sin\end{matrix} \frac{2\pi m}{D} x \tag{11.1}$$

in a one-dimensional case, where D is the actual length which is taken as 2π in the Fourier series. Differentiating this with respect to x, the horizontal gradient of gravity is expressed by

$$\frac{\partial g}{\partial x} = \frac{\partial^2 V}{\partial z\, \partial x} = \sum_m \mp \frac{2\pi m}{D} B_m \begin{matrix}\sin\\ \cos\end{matrix} \frac{2\pi m}{D} x.$$

The coefficient $2\pi m/D$ on the right-hand side of the above equation increases with m. This means that $\partial^2 V/\partial z\, \partial x$ is more sensitive to variations with larger m (shorter wavelength component) than to those with smaller m (longer wavelength component). Underground geological structures which are responsible for $\partial^2 V/\partial z\, \partial x$ and which have larger m generally lie at shallower depths. This is one of the reasons why second derivatives of gravity potential are useful for finding small geological structures, which are often associated with natural resources such as oil and coal.

In Figure 11.1 the curve $(\sin x + \frac{1}{3}\sin 3x)$ and its derivative $(\cos x + \cos 3x)$ are compared graphically. It is seen that, by differentiation, the shorter wavelength component becomes more enhanced than the longer one.

11.2 Curvature of the geoid

Let us see how the three second derivatives of gravity potential, $\partial^2 V/\partial x^2$, $\partial^2 V/\partial x\, \partial y$ and $\partial^2 V/\partial y^2$, are related to the curvature of the geoid. Gravity potential V in the neighbourhood of the origin can be expressed in a Taylor's series as follows:

$$V(xyz) = V(0) + \left(\frac{\partial V}{\partial x}\right)x + \left(\frac{\partial V}{\partial y}\right)y + \left(\frac{\partial V}{\partial z}\right)z$$
$$+ \frac{1}{2}\left(\frac{\partial^2 V}{\partial x^2}\right)x^2 + \frac{1}{2}\left(\frac{\partial^2 V}{\partial y^2}\right)y^2 + \frac{1}{2}\left(\frac{\partial^2 V}{\partial z^2}\right)z^2$$
$$+ \left(\frac{\partial^2 V}{\partial x\, \partial y}\right)xy + \left(\frac{\partial^2 V}{\partial y\, \partial z}\right)yz + \left(\frac{\partial^2 V}{\partial z\, \partial x}\right)zx + \ldots . \tag{11.2}$$

On the equipotential surface that passes the origin, $V(xyz)$ has a constant value $V(0)$, so that

$$V(xyz) = V(0).$$

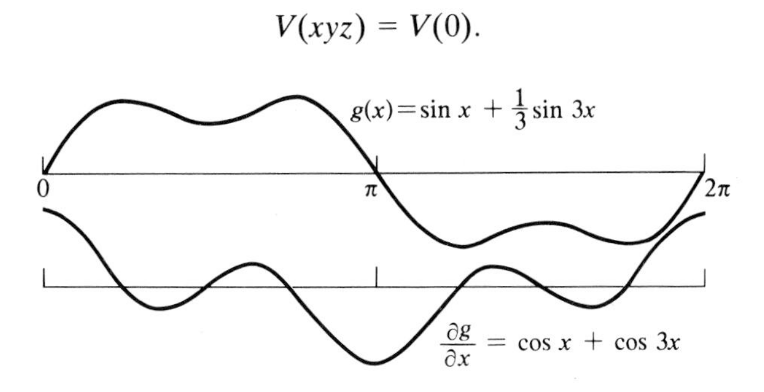

Figure 11.1 Comparison of $g(x) = \sin x + \frac{1}{3}\sin 3x$ and $\partial g/\partial x = \cos x + \cos 3x$.

Since the z-axis is taken vertically downwards,

$$\frac{\partial V}{\partial x} = 0 \qquad \frac{\partial V}{\partial y} = 0$$

at the origin. Also, since the equipotential surface is very flat and does not differ much from a geometrical plane, z is very small compared with x and y,

$$z \ll x \qquad z \ll y.$$

With these approximations, the expression for the equipotential surface becomes

$$gz + \frac{1}{2}\left(\frac{\partial^2 V}{\partial x^2}x^2 + \frac{\partial^2 V}{\partial y^2}y^2\right) + \frac{\partial^2 V}{\partial x\,\partial y}xy = 0. \qquad (11.3)$$

Suppose that this equipotential surface is cut by a vertical plane which includes the z-axis and which makes an angle α with the x-axis as shown in Figure 11.2. The intersection of this vertical plane with the equipotential surface is a circular arc in the neighbourhood of the origin. The radius R of this arc is the radius of curvature of the equipotential surface in the direction α and is given by

$$R = g\bigg/\left(\frac{\partial^2 V}{\partial x^2}\cos^2\alpha + \frac{\partial^2 V}{\partial y^2}\sin^2\alpha + 2\frac{\partial^2 V}{\partial x\,\partial y}\sin\alpha\cos\alpha\right). \qquad (11.4)$$

R changes according to the direction α of the vertical plane. R is smallest or largest when the denominator on the right-hand side of Equation (11.4) for R is largest or smallest, respectively. The value of α that makes the denominator largest or smallest can be found by differentiating it with respect to α and putting the result equal to zero:

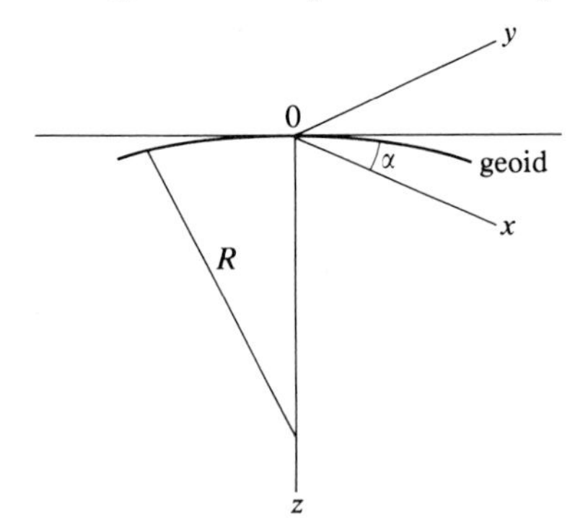

Figure 11.2 Intersection of the geoid and a vertical plane.

$$\frac{\partial}{\partial \alpha}\left(\frac{\partial^2 V}{\partial x^2}\cos^2\alpha + \frac{\partial^2 V}{\partial y^2}\sin^2\alpha + 2\frac{\partial^2 V}{\partial x\, \partial y}\sin\alpha\cos\alpha\right) = 0.$$

From this, it follows that

$$\tan 2\alpha = -2\frac{\partial^2 V}{\partial x\, \partial y}\bigg/\left(\frac{\partial^2 V}{\partial y^2} - \frac{\partial^2 V}{\partial x^2}\right).$$

This equation gives two values, α_1 and α_2, which are the two directions corresponding to the maximum and the minimum curvatures, $1/R_1$ and $1/R_2$, and which differ by $\pi/2$. The maximum and minimum curvatures are

$$\frac{1}{R_1} = -\frac{1}{2g}\left\{\left(\frac{\partial^2 V}{\partial x^2} + \frac{\partial^2 V}{\partial y^2}\right) + \left(\frac{\partial^2 V}{\partial y^2} - \frac{\partial^2 V}{\partial x^2}\right)\sec 2\alpha_1\right\},$$

$$\frac{1}{R_2} = \frac{1}{2g}\left\{\left(\frac{\partial^2 V}{\partial x^2} + \frac{\partial^2 V}{\partial y^2}\right) + \left(\frac{\partial^2 V}{\partial y^2} - \frac{\partial^2 V}{\partial x^2}\right)\sec 2\alpha_2\right\},$$

and the difference of the two is

$$\frac{1}{R_1} - \frac{1}{R_2} = \frac{1}{g}\sqrt{\left\{\left(\frac{\partial^2 V}{\partial y^2} - \frac{\partial^2 V}{\partial x^2}\right)^2 + 4\left(\frac{\partial^2 V}{\partial x\, \partial y}\right)^2\right\}}.$$

As seen in the above, the second derivatives, $\partial^2 V/\partial x^2$, $\partial^2 V/\partial x\, \partial y$, $\partial^2 V/\partial y^2$ are related to the directions and the difference of the maximum and minimum curvatures of the geoid. It is because of this that they are called the curvature terms.

11.3 Horizontal gradient of gravity

The second derivatives $\partial^2 V/\partial z\, \partial x$ and $\partial^2 V/\partial z\, \partial y$ are the x- and y-components of the rate of horizontal variation of g:

$$\frac{\partial^2 V}{\partial z\, \partial x} = \frac{\partial g}{\partial x}$$

$$\frac{\partial^2 V}{\partial z\, \partial y} = \frac{\partial g}{\partial y}.$$

The maximum gradient of g is given by

$$G = \sqrt{\left\{\left(\frac{\partial g}{\partial x}\right)^2 + \left(\frac{\partial g}{\partial y}\right)^2\right\}}$$

$$= \sqrt{\left\{\left(\frac{\partial^2 V}{\partial z\, \partial x}\right)^2 + \left(\frac{\partial^2 V}{\partial z\, \partial y}\right)^2\right\}} \tag{11.5}$$

and its direction by

$$\tan\theta = \frac{\partial g}{\partial y}\Big/\frac{\partial g}{\partial x}$$

$$= \frac{\partial^2 V}{\partial z\, \partial y}\Big/\frac{\partial^2 V}{\partial z\, \partial x}. \tag{11.6}$$

$\partial^2 V/\partial z\, \partial x$ and $\partial^2 V/\partial z\, \partial y$ are called the gradient terms.

11.4 Magnitude of the second derivatives

Let us estimate the magnitude of the second derivatives by means of a simple model. Suppose, for instance, there is an infinitely long straight mass of a linear density σ extending in the direction perpendicular to the plane of paper at a depth d below the ground surface. The component of attraction in the x-direction due to this mass is

$$X = \frac{\partial V}{\partial x} = -2G\sigma\frac{x}{x^2 + d^2} \tag{11.7}$$

and its derivative in the direction of x is

$$\frac{\partial X}{\partial x} = \frac{\partial^2 V}{\partial x^2} = -2G\sigma\frac{d^2 - x^2}{(x^2 + d^2)^2}.$$

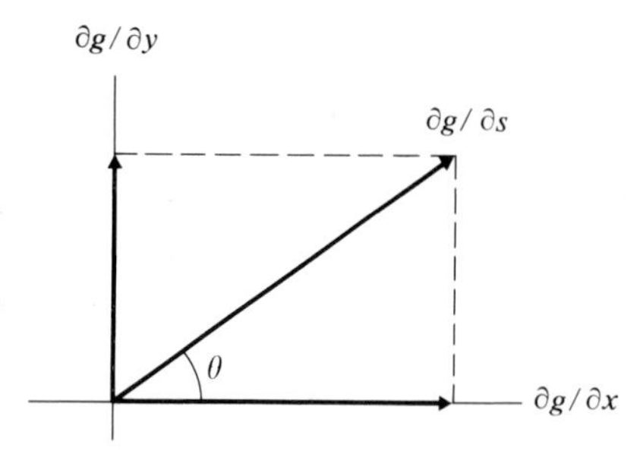

Figure 11.3 Maximum gravity gradient G.

This is a maximum and a minimum at $x = \pm\sqrt{3}d$ and at $x = 0$ respectively. The value of $\partial^2 V/\partial x^2$ is $-2G\sigma/d^2$ at $x = 0$ and is $G\sigma/4d^2$ at $x = \pm\sqrt{3}d$. If, for instance, the mass is 0.3 higher in density than its surroundings and has a cross-sectional area of 10^{10} cm² and a depth of 5×10^5 cm, then $\partial^2 V/\partial x^2$ at $x = 0$ is approximately

$$-\frac{2G\sigma}{d^2} = -\frac{2 \times 6.67 \times 10^{-8} \times 0.3 \times 10^{10}}{(5 \times 10^5)^2} \approx -1.6 \times 10^{-9}.$$

The second derivatives are generally several tens or hundreds in units of 10^{-9}. This unit of 10^{-9} is called one Eötvös in memory of R. v. Eötvös (1847–1919) who first pointed out the significance of second derivatives in geophysical problems and invented an instrument to measure them.

11.5 The gravity variometer or torsion balance

An instrument by which second derivatives of gravity potential can be measured directly was invented by R. v. Eötvös and it is called a gravity variometer or a torsion balance. The instrument is very simple in principle. A rod, which has two equal masses m attached to both its ends is suspended horizontally from above as shown in Figure 11.4. The direction of gravity at the point O is taken as the z-axis. Going into high accuracies in subsequent discussions, the gravity forces which are acting on the two end masses are not geometrically parallel to the z-axis and have horizontal components g_x and g_y working in the x- and y-directions respectively. As a result, a rotational moment of forces

$$M = \int (g_y x - g_x y)\,\mathrm{d}m \tag{11.8}$$

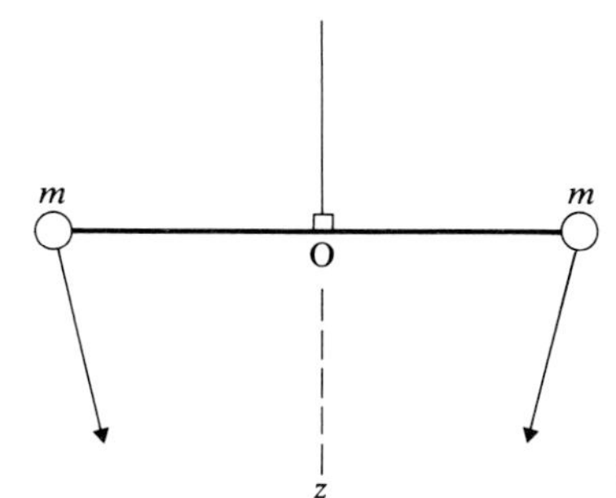

Figure 11.4 A torsion pendulum.

works on the rod around the z-axis. By Taylor's series expansion, g_x and g_y are given by

$$g_x = \frac{\partial V}{\partial x} = \left(\frac{\partial V}{\partial x}\right)_0 + \frac{\partial^2 V}{\partial x^2} x + \frac{\partial^2 V}{\partial x\,\partial y} y = \frac{\partial^2 V}{\partial x^2} x + \frac{\partial^2 V}{\partial x \partial y} y,$$

$$g_y = \frac{\partial V}{\partial y} = \left(\frac{\partial V}{\partial y}\right)_0 + \frac{\partial^2 V}{\partial x\,\partial y} x + \frac{\partial^2 V}{\partial y^2} y = \frac{\partial^2 V}{\partial x\,\partial y} x + \frac{\partial^2 V}{\partial y^2} y,$$

because

$$\left(\frac{\partial V}{\partial x}\right)_0 = \left(\frac{\partial V}{\partial y}\right)_0 = 0.$$

When these are put in the expression of the moment M, it reduces to

$$M = \left(\frac{\partial^2 V}{\partial y^2} - \frac{\partial^2 V}{\partial x^2}\right)\int xy\ \mathrm{d}m + \frac{\partial^2 V}{\partial x\,\partial y}\int (x^2 - y^2)\ \mathrm{d}m. \qquad (11.9)$$

The x- and y-co-ordinates used here in the integrals refer to the axes fixed in space. Now these will be transformed into ξ- and η-co-ordinates that refer to the rod. The ξ-axis is taken in the direction of the length of the rod and the η-axis perpendicular to it. The transformation between (x, y) and (ξ, η) is

$$x = \xi \cos\alpha - \eta \sin\alpha,$$

$$y = \xi \sin\alpha + \eta \cos\alpha,$$

where α is the angle between the ξ-axis and the x-axis. Then we have

$$\int xy\ \mathrm{d}m = \frac{\sin 2\alpha}{2}\int (\xi^2 - \eta^2)\ \mathrm{d}m + \cos 2\alpha \int \xi\eta\ \mathrm{d}m,$$

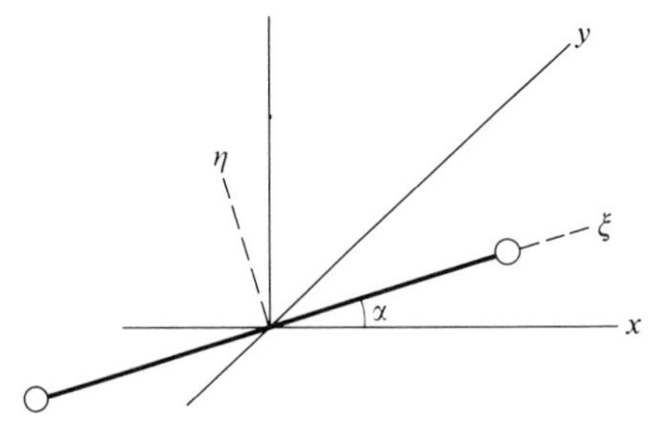

Figure 11.5 Transformations between (x, y) and (ξ, η).

$$\int (x^2 - y^2)\, dm = \cos 2\alpha \int (\xi^2 - \eta^2)\, dm - 2 \sin 2\alpha \int \xi\eta\, dm,$$

and putting these into the expression of the moment M, we obtain

$$M = \left(\frac{\partial^2 V}{\partial y^2} - \frac{\partial^2 V}{\partial x^2}\right)\left\{\frac{\sin 2\alpha}{2}\int (\xi^2 - \eta^2)\, dm + \cos 2\alpha \int \xi\eta\, dm\right\}$$

$$+ \frac{\partial^2 V}{\partial x\, \partial y}\left\{\cos 2\alpha \int (\xi^2 - \eta^2)\, dm - 2 \sin 2\alpha \int \xi\eta\, dm\right\}. \quad (11.10)$$

The rod is long in the direction of ξ and thin in the direction perpendicular to it, so that $\int (\xi^2 - \eta^2)\, dm$ is nearly equal to $\int (\xi^2 + \eta^2)\, dm$ which is the moment of inertia I of the rod around the z-axis. Also

$$\int \xi\eta\, dm = 0$$

by symmetry. The moment M is finally

$$M = \frac{I}{2}\left(\frac{\partial^2 V}{\partial y^2} - \frac{\partial^2 V}{\partial x^2}\right)\sin 2\alpha + I\frac{\partial^2 V}{\partial x\, \partial y}\cos 2\alpha. \quad (11.11)$$

By this moment M, the rod will be deflected from the zero position by a small angle θ. This angle of deflection θ is where the elastic torque $\tau\theta$ produced in the suspension wire will balance with M. Thus

$$\tau\theta = \frac{I}{2}\left(\frac{\partial^2 V}{\partial y^2} - \frac{\partial^2 V}{\partial x^2}\right)\sin 2\alpha + I\frac{\partial^2 V}{\partial x\, \partial y}\cos 2\alpha. \quad (11.12)$$

In this equation, $(\partial^2 V/\partial y^2) - (\partial^2 V/\partial x^2)$ and $\partial^2 V/\partial x\, \partial y$ are unknown, while τ are I are known instrumental constants. As to θ, it is measured from the undeflected position θ_0 of the rod which is unknown. The above equation of equilibrium should therefore be written as

$$\tau(\theta - \theta_0) = \frac{I}{2}\left(\frac{\partial^2 V}{\partial y^2} - \frac{\partial^2 V}{\partial x^2}\right)\sin 2\alpha + I\frac{\partial^2 V}{\partial x\, \partial y}\cos 2\alpha$$

with three unknowns, θ_0, $(\partial^2 V/\partial y^2) - (\partial^2 V/\partial x^2)$ and $\partial^2 V/\partial x\, \partial y$.

Now suppose that the instrument is set at three different angular positions successively so that α will be α_1, α_2, α_3, and the corresponding angles of the twist of the wire are $(\theta_1 - \theta_0)$, $(\theta_2 - \theta_0)$, and $(\theta_3 - \theta_0)$. When the whole instrument is turned through a certain angle from one setting to another, the suspended rod does not turn by the same amount.

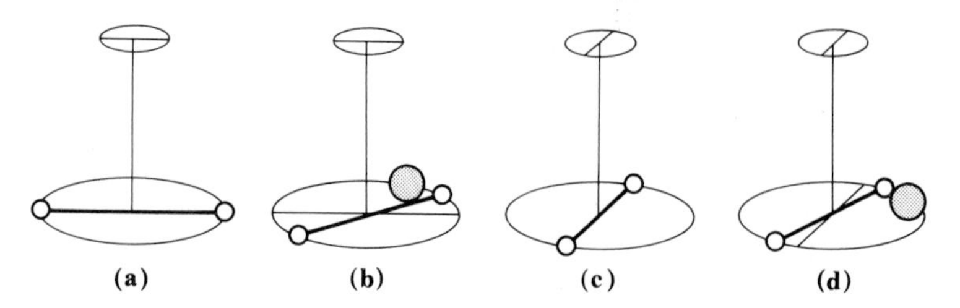

Figure 11.6 Equilibrium positions of a torsion pendulum when there is a mass (b, d) and when there is none (a, c).

This circumstance will be made clear by an illustration given in Figure 11.6. If there is no mass as in Figure 11.6a and c and if the head of the wire is turned through 90°, for instance, the rod will also turn through the same angle of 90°. But in the case when there is a big mass M near the instrument (Fig. 11.6b & d), if the head of the wire is turned through 90°, the rod will not turn through the same angle of 90°. There is a small difference in turning angles of the head of the wire and the rod. It is from these differences that the second derivatives of gravity potential can be found.

If we taken 0°, 120°, 240° for α_1, α_2, α_3, the equations of equilibrium of the rod at the three settings become

$$\tau(\theta_1 - \theta_0) = I\frac{\partial^2 V}{\partial x\,\partial y},$$

$$\tau(\theta_2 - \theta_0) = -\frac{\sqrt{3}}{4}I\left(\frac{\partial^2 V}{\partial y^2} - \frac{\partial^2 V}{\partial x^2}\right) - \frac{I}{2}\frac{\partial^2 V}{\partial x\,\partial y},$$

$$\tau(\theta_3 - \theta_0) = \frac{\sqrt{3}}{4}I\left(\frac{\partial^2 V}{\partial y^2} - \frac{\partial^2 V}{\partial x^2}\right) - \frac{I}{2}\frac{\partial^2 V}{\partial x\,\partial y}.$$

Eliminating θ_0 from these equations, we obtain

$$\tau(\theta_1 - \theta_2) = \frac{\sqrt{3}}{4}I\left(\frac{\partial^2 V}{\partial y^2} - \frac{\partial^2 V}{\partial x^2}\right) + \frac{3}{2}I\frac{\partial^2 V}{\partial x\,\partial y},$$

$$\tau(\theta_1 - \theta_3) = -\frac{\sqrt{3}}{4}I\left(\frac{\partial^2 V}{\partial y^2} - \frac{\partial^2 V}{\partial x^2}\right) + \frac{3}{2}I\frac{\partial^2 V}{\partial x\,\partial y},$$

and the second derivatives can be found by

$$\frac{\partial^2 V}{\partial y^2} - \frac{\partial^2 V}{\partial x^2} = \frac{2\tau}{\sqrt{3}I}\{(\theta_1 - \theta_2) - (\theta_1 - \theta_3)\},$$

$$\frac{\partial^2 V}{\partial x\, \partial y} = \frac{\tau}{3I}\{(\theta_1 - \theta_2) + (\theta_1 - \theta_3)\}.$$

θ_1, θ_2 and θ_3 are different and $(\theta_1 - \theta_2)$ and $(\theta_1 - \theta_3)$ are the differences of deflection angles at the three settings α_1, α_2, α_3.

Since the second derivatives of potential to be measured are of the order of 10^{-9}, the magnitude θ of the angle of deflection of the rod that has to be observed is of the order of $I \times 10^{-9}/\tau$. This can be done by the telescope and scale method, visually or photographically. The telescope and the scale are to be fixed to the instrument. Rotate the whole instrument through a certain angle after the suspended rod becomes stationary at one setting. Different scale values θ will be seen in the telescope. In order to make this difference measurable with an accuracy of at least 10^{-4} in angle, I/τ must be

$$10^{-4} = \frac{I}{\tau} \times 10^{-9}$$

$$\frac{I}{\tau} = 10^5.$$

For this purpose, I must be made large and τ small. In that case, the period T of free-rotational oscillation of the rod will be as long as

$$T = 2\pi\sqrt{\left(\frac{I}{\tau}\right)} \approx 6 \times \sqrt{10^5} \approx 2000 \text{ s} \approx 30 \text{ min}.$$

Such a long-period pendulum is easily disturbed by air currents around it and therefore it must be carefully protected from the outside by means of thermal shieldings. About one hour will be needed before the rod comes to rest, so that observations at three different settings will take roughly five hours altogether.

If the suspended pendulum has a form as shown in Figure 11.7a, the equation of its equilibrium is

$$\theta - \theta_0 = a\left(\frac{\partial^2 V}{\partial y^2} - \frac{\partial^2 V}{\partial x^2}\right)\sin 2\alpha + 2a\frac{\partial^2 V}{\partial x\, \partial y}\cos 2\alpha - b\frac{\partial^2 V}{\partial z\, \partial x}\sin\alpha + b\frac{\partial^2 V}{\partial y\, \partial z}\cos\alpha \quad (11.13)$$

where a and b are instrumental constants. In this case, there are five unknowns

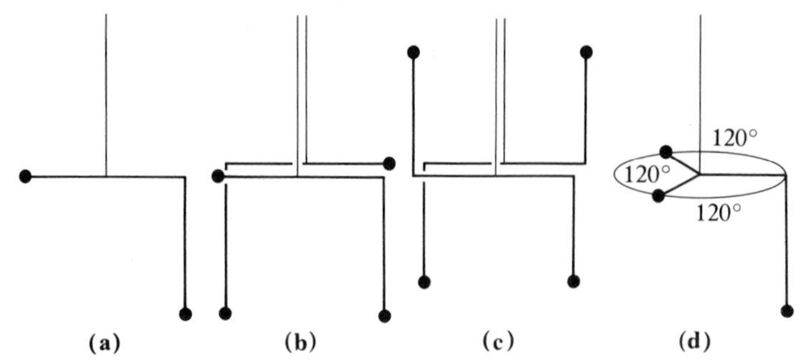

Figure 11.7 Various types of torsion pendulum.

$$\theta_0 \qquad \left(\frac{\partial^2 V}{\partial y^2} - \frac{\partial^2 V}{\partial x^2}\right) \qquad \frac{\partial^2 V}{\partial x\, \partial y} \qquad \frac{\partial^2 V}{\partial z\, \partial x} \qquad \frac{\partial^2 V}{\partial y\, \partial z},$$

so that observations at five different settings are needed. In order to shorten the time of observation, there is a device to suspend two similar pendulums facing in opposite directions as shown in Figure 11.7b. In this case, observations in three different settings are sufficient. By making the pendulum in a form as shown in Figure 11.7c, the whole instrument can be made smaller in size. If the pendulum has a form as shown in Figure 11.7d, the equation of equilibrium is

$$\theta - \theta_0 = b \frac{\partial^2 V}{\partial y\, \partial z} \cos \alpha - b \frac{\partial^2 V}{\partial z\, \partial x} \sin \alpha.$$

From this, the horizontal gradients of gravity, $\partial g/\partial x$ and $\partial g/\partial y$ only can be known by observations at three different settings. In Figure 11.8 two models of torsion balance are shown.

11.6 Graphical representations

In order to show the second derivatives found by a torsion balance, graphical representations as in Figures 11.3 and 11.9 can be conveniently made.

Take $\partial g/\partial x$ on the x-axis and $\partial g/\partial y$ on the y-axis and compose them vectorially. The resultant gives the direction and magnitude of the largest horizontal gradient of gravity. To show the curvature terms:

(a) take $-2(\partial^2 V/\partial x\, \partial y)$ along the x-axis;
(b) take $(\partial^2 V/\partial y^2) - (\partial^2 V/\partial x^2)$ along the y-axis;
(c) draw the diagonal OC;
(d) draw the bisector of the angle BOC;
(e) take OD = OE = OC/2 along the bisector.

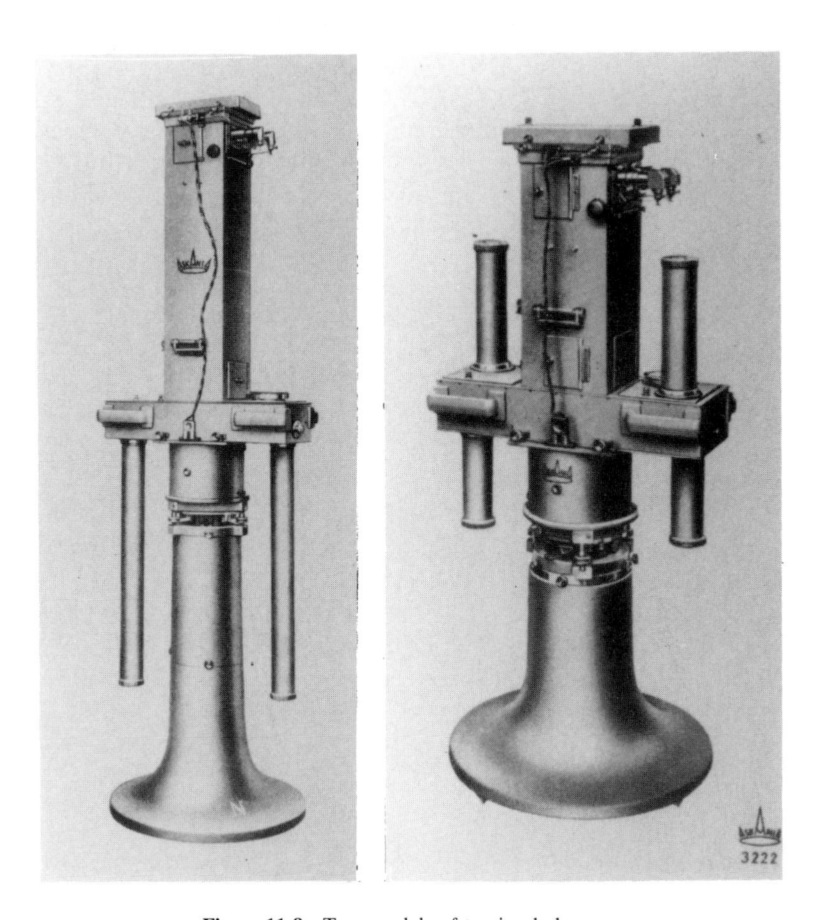

Figure 11.8 Two models of torsion balance.

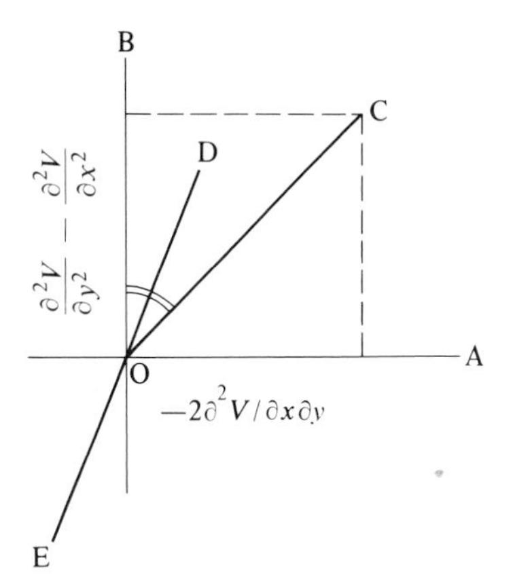

Figure 11.9 Graphical representation of curvature terms.

The segment DE gives the direction of maximum curvature and its length represents $g(1/R_1) - (1/R_2)$. The above two combined, as in Figure 11.10, will give us sufficient information concerning the second derivatives of gravity potential at one point. An example of their geographical distribution is shown in Figure 11.11. This was observed by T. Minakami around Asama volcano, Japan, in 1937.

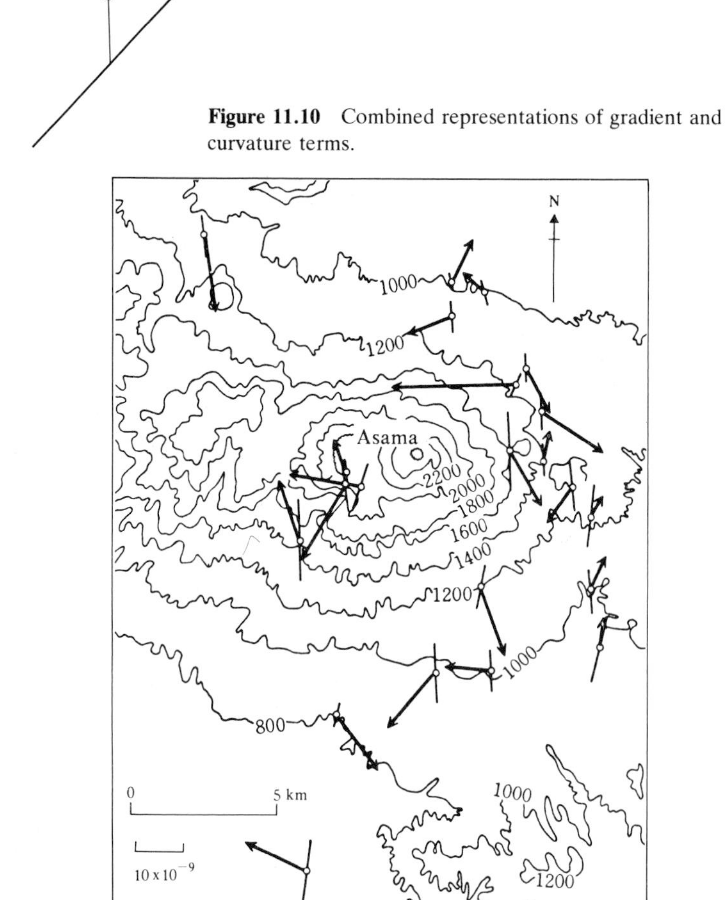

Figure 11.10 Combined representations of gradient and curvature terms.

Figure 11.11 Geographical distribution of second derivatives around Asama volcano, Japan.

11.7 Interpretations

If the distribution of second derivatives is known, the next problem is how to find the responsible underground mass distributions. If a semi-infinite plane mass is considered to extend from $\xi = 0$ to $\xi = \infty$ at a depth d as shown in Figure 11.12;

(a) $$\frac{\partial V}{\partial z} = g \propto \frac{\pi}{2} + \tan^{-1}\frac{x}{d};$$

(b) $$\frac{\partial^2 V}{\partial z\,\partial x} = \frac{\partial g}{\partial x} \propto \frac{d}{x^2 + d^2};$$

(c) $$\frac{\partial^2 V}{\partial x^2} = \propto - \frac{x}{x^2 + d^2}.$$

Three curves for (a), (b) and (c) are shown in the same figure. As are clearly seen in the figure, the curves for the second derivatives, (b) and (c), are much sharper than the curve (a) for the first derivative. This is a good example indicating that the second-derivative distribution is useful for locating underground discontinuities accurately.

If an underground-mass distribution is once assumed, it is not difficult to calculate the resulting distribution of second derivatives. But it is not an easy thing to go in the other direction. For doing this, the Fourier series method which was introduced before is again useful.

Suppose there is a plane mass distribution

$$M(x) = \sum C_m \begin{matrix}\cos\\ \sin\end{matrix} \frac{2\pi m}{D} x \tag{11.14}$$

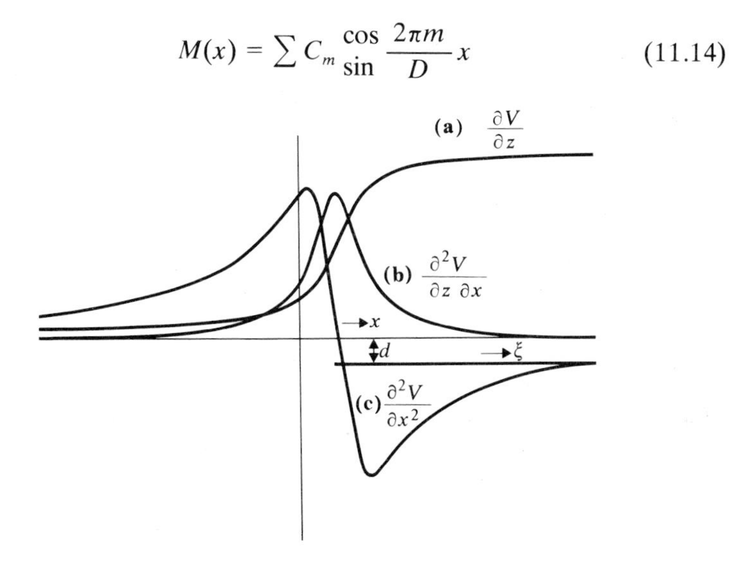

Figure 11.12 Distribution of $\partial V/\partial z$, $\partial^2 V/\partial z\,\partial x$ and $\partial^2 V/\partial x^2$ due to a semi-infinite plane mass.

at a depth d. Then, the distribution of gravity on the Earth's surface is

$$g(x) = 2\pi G \sum_m C_m \begin{matrix}\cos\\ \sin\end{matrix} \frac{2\pi m}{D} x \exp\left(-\frac{2\pi m}{D} d\right)$$

and that of $\partial g/\partial x$ is

$$\frac{\partial g}{\partial x} = \mp 2\pi G \sum_m \frac{2\pi m}{D} C_m \begin{matrix}\sin\\ \cos\end{matrix} \frac{2\pi m}{D} x \exp\left(-\frac{2\pi m}{D} d\right).$$

Conversely, if

$$\frac{\partial g}{\partial x} = \sum_m G_m \begin{matrix}\cos\\ \sin\end{matrix} \frac{2\pi m}{D} x,$$

the responsible underground mass can at once be found from

$$M(x) = \mp \frac{1}{2\pi G} \sum_m \frac{D}{2\pi m} G_m \begin{matrix}\sin\\ \cos\end{matrix} \frac{2\pi m}{D} x \exp\left(\frac{2\pi m}{D} d\right).$$

In words, expand the distribution of $\partial g/\partial x$ in a Fourier series, multiply the Fourier coefficients by

$$\frac{D}{2\pi m} \exp\left(\frac{2\pi m}{D} d\right)$$

and synthesize a series of interchanging cosine and sine terms.

For demonstrating the use of this method, it will be applied to the $\partial g/\partial x$ distribution calculated by Eötvös in his original paper. He took a model, as shown in Figure 11.13 in which a layer with a density ρ overlies a layer of a density $(\rho + 0.8)$ with an irregular interface. It is difficult to find the shape of this irregular interface from the distribution of $\partial g/\partial x$ along the Earth's surface without the help of the Fourier series method. The depth d of the underground mass will be taken to be 551 m following the original model of Eötvös. After the mass distribution $M(x)$ is found by the Fourier series method mentioned above, it is divided by 0.8, which is the difference in density between the two layers. The result will give the form of their interface. The shape of the interface so calculated is shown by dots in Figure 11.13. It is seen that the shape of the calculated interface agrees satisfactorily with the original model of Eötvös.

Torsion-balance observations have been very useful in finding local shallow geological structures. Recent progress of static gravimeters,

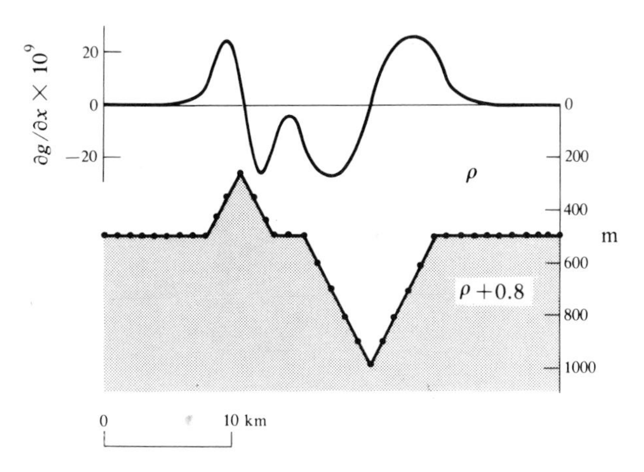

Figure 11.13 Eötvös' model of an underground structure and calculated distribution of $\partial g/\partial x$. Shape of the irregular interface calculated by the Fourier series method.

however, has made it possible for us to measure gravity values with an accuracy of 0.01 mgal. If these measurements are made at a distance of 100 m, the horizontal gradient of gravity can be known with an accuracy of one Eötvös (10^{-9}) within a much shorter time than by a torsion balance;

$$0.01 \times 10^{-3} \div 10^{4} = 10^{-9}.$$

In this sense, the gravimeter may be said to have replaced the position of the torsion balance.

11.8 Vertical gradient of gravity

The change in gravity value according to height is $\partial g/\partial z$, or $\partial^2 V/\partial z^2$, which is the second derivative of potential with respect to z. Usually the value of 3086×10^{-9} is taken and is used for calculating the free-air gravity correction. This value is obtained for the average Earth. But strictly, the value varies from one place to another.

Suppose that there are two points A and B on the ground and that g_{A} is larger than g_{B}. We know that if we go to a height h, the difference in gravity values at A′ and B′ will become smaller,

$$g_{\mathrm{A}} - g_{\mathrm{B}} > g_{\mathrm{A'}} - g_{\mathrm{B'}}$$

or

$$g_{\mathrm{A}} - g_{\mathrm{A'}} > g_{\mathrm{B}} - g_{\mathrm{B'}};$$

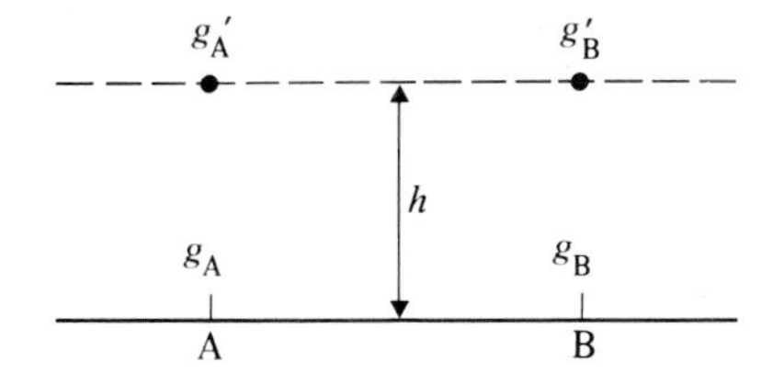

Figure 11.14 g_A and g_B at $z = 0$ and g_A' and g_B' at $z = h$.

therefore

$$\frac{g_A - g_{A'}}{h} > \frac{g_B - g_{B'}}{h}.$$

This means that the vertical gradient of gravity is larger at a point where g is large than at a point where g is small. Since gravity potential V is

$$V = \sum_m A_m \begin{matrix}\cos\\ \sin\end{matrix} \frac{2\pi m}{D} x \exp\left(\frac{2\pi m}{D} z\right), \tag{11.15}$$

the gravity value and its vertical gradient on the Earth's surface $z = 0$ are

$$\begin{aligned} g &= \frac{\partial V}{\partial z} \\ &= \sum_m \frac{2\pi m}{D} A_m \begin{matrix}\cos\\ \sin\end{matrix} \frac{2\pi m}{D} x \\ &= \sum_m B_m \begin{matrix}\cos\\ \sin\end{matrix} \frac{2\pi m}{D} x; \end{aligned}$$

$$\begin{aligned} \frac{\partial g}{\partial z} &= \frac{\partial^2 V}{\partial z^2} \\ &= \sum_m \left(\frac{2\pi m}{D}\right)^2 A_m \begin{matrix}\cos\\ \sin\end{matrix} \frac{2\pi m}{D} x \\ &= \sum_m \frac{2\pi m}{D} B_m \begin{matrix}\cos\\ \sin\end{matrix} \frac{2\pi m}{D} x. \end{aligned}$$

From these expressions, it is seen that the Fourier components for g and for $\partial g/\partial z$ are in phase with each other and the Fourier coefficients for $\partial g/\partial z$ are $2\pi m/D$ times those for g. $2\pi m/D$ is equal to $2\pi/\lambda_m$ where λ_m is the wavelength of a component m. If there is a harmonic variation of g with an amplitude B_m, the amplitude of variation of $\partial g/\partial z$ is $(2\pi/\lambda_m)B_m$. Table 11.1 gives the values of $(2\pi/\lambda_m)B_m$ for various B_m and λ_m.

Table 11.1 Values for $(2\pi/\lambda_m)B_m$ in 10^{-9}.

$\lambda m \backslash Bm$	1 mgal	5 mgal	10 mgal
1 km	63	314	628
5 km	13	63	126
10 km	6	31	63

Anomalies in $\partial g/\partial z$ cannot be neglected when compared with its normal value of 3086×10^{-9} especially when B_m is large and λ_m small.

The Fourier–Bessel series can also be conveniently applied for calculating anomalies in the vertical gradient of gravity at a point from the distribution of gravity anomalies around that point. The general expression of gravity potential at $z = z$ in cylindrical co-ordinates has already been shown to be

$$V = \sum_n \int_0^\infty \{A_{nk} J_n(kr)\, \mathrm{d}k\} \begin{matrix}\cos\\ \sin\end{matrix}\, n\theta\, \mathrm{e}^{-kz} \tag{11.16}$$

and that of gravity at $z = 0$ to be

$$\begin{aligned} g &= \sum_n \int_0^\infty \{A_{nk} k J_n(kr)\, \mathrm{d}k\} \begin{matrix}\cos\\ \sin\end{matrix}\, n\theta \\ &= \sum_n \int_0^\infty \{B_{nk} J_n(kr)\, \mathrm{d}k\} \begin{matrix}\cos\\ \sin\end{matrix}\, n\theta \\ &= \int_0^\infty B_{0k} J_0(kr)\, \mathrm{d}k + \left\{\int_0^\infty B_{1k} J_1(kr)\, \mathrm{d}k\right\} \begin{matrix}\cos\\ \sin\end{matrix}\, \theta \\ &\quad + \left\{\int_0^\infty B_{2k} J_2(kr)\, \mathrm{d}k\right\} \begin{matrix}\cos\\ \sin\end{matrix}\, 2\theta + \ldots . \end{aligned}$$

The anomaly of vertical gradient $\partial g/\partial z$ at $r = 0$ is therefore

$$\frac{\partial g}{\partial z} = \int_0^\infty k B_{0k}\, \mathrm{d}k,$$

remembering $J_0(0) = 1$, $J_1(0) = J_2(0) = \ldots = 0$. By using a theorem about a Bessel function

$$\int_0^\infty \frac{\{1 - J_0(kr)\}}{r^2}\, \mathrm{d}r = k$$

and a little further mathematics, it is seen that the right-hand side of the above equation is transformed into

$$\frac{\partial g}{\partial z} = \int_0^\infty kB_{0k}\, \mathrm{d}k = \frac{\overline{g(0)} - \overline{g(R)}}{R} + \int_R^\infty \frac{\overline{g(0)} - \overline{g(r)}}{r^2}\, \mathrm{d}r,$$

where $\overline{g(r)}$ is the mean value of g taken along a circle $r = r$ and R is the limit of r up to which $\{\overline{g(0)} - \overline{g(r)}\}$ can be regarded as changing accord- to αr^2.

Figure 11.15 shows the variations with r of

$$\{\overline{g(0)} - \overline{g(r)}\} \qquad \frac{\overline{g(0)} - \overline{g(r)}}{r^2} \qquad \int_R^\infty \frac{\overline{g(0)} - \overline{g(r)}}{r^2}\, \mathrm{d}r$$

at a point $\varphi = 34.3°\mathrm{N}$, $\lambda = 98.3°\mathrm{W}$ in the USA. In the figure, $\{\overline{g(0)} - \overline{g(r)}\}$ is seen to vary parabolically up to $R = 40$ km where the value is 63 mgal. The term

$$\int_R^t \frac{\overline{g(0)} - \overline{g(r)}}{r^2}\, \mathrm{d}r$$

tends to 21×10^{-9}. Thus finally

$$\frac{\partial g}{\partial z} = \frac{63 \times 10^{-3}}{40 \times 10^5} + 21 \times 10^{-9} = 37 \times 10^{-9}.$$

In this example, $\partial g/\partial z$ is a little more than 1% larger than the normal value, 3086×10^{-9}.

The vertical gradient of gravity can be found directly by measuring the gravity value at the top of a high tower or a building and comparing it with the value on the ground. The difference in gravity when divided by the height of the tower or the building gives the vertical gradient. The values of $\partial g/\partial z$ at various places have been measured by this method. It is not unusual that $\partial g/\partial z$ values differ from the normal value by 10%.

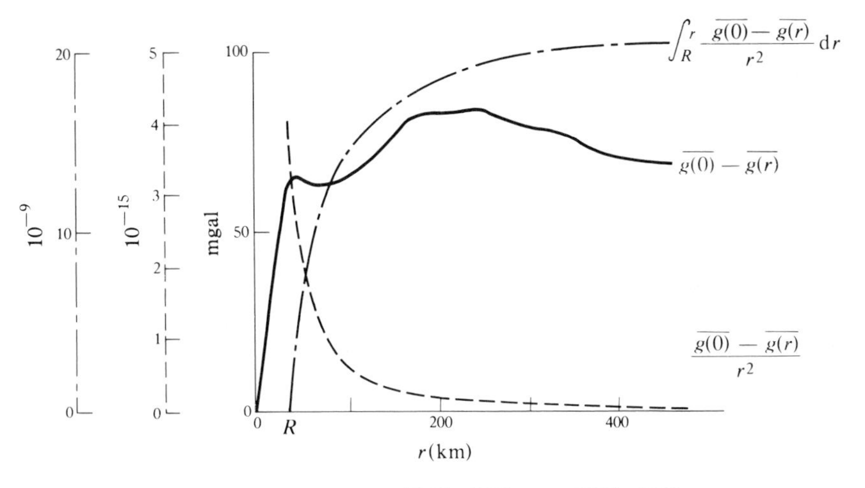

Figure 11.15 Variations of $\{\overline{g(0)} - \overline{g(r)}\}$, $\frac{\overline{g(0)} - \overline{g(r)}}{r^2}$ and $\int_R^r \frac{\overline{g(0)} - \overline{g(r)}}{r^2} dr$ according to r.

11.9 Remarks on free-air reduction

If $\partial g/\partial z$ at a point differs from its normal value, 3086×10^{-9}, the usual free-air reduction using this value will not give a correct reduction. If, for instance, $\partial g/\partial z$ at a point is 10% different from the normal value, this will make a difference of 30 mgal for a height of 1000 m.

Referring to Figure 11.16, the normal free-air reduction means drawing a straight line with an inclination of 3086×10^{-9} from the point G where the gravity value is g and to find g_0 at a point at which this straight line crosses the y-axis. If, as in Figure 11.16, the value of g_0 so reduced is larger than the standard gravity value γ for that point, the free-air anomaly is positive and therefore the inclination of the straight line should have been steeper as shown by the dotted line. Then, the difference of g_0 and γ will be larger and the inclination should be still larger and so on. This is unexpected. Such a situation as this arises from too

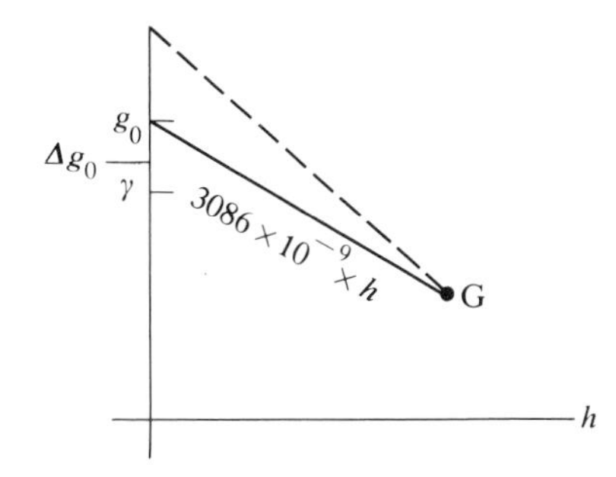

Figure 11.16 Graphical explanation of free-air reduction.

simple an assumption that $\partial g/\partial z$ is constant between the point G and the geoid. Since the free-air anomaly is

$$\begin{aligned} & g + (\text{normal free-air reduction}) - (\text{normal gravity at 0}) \\ & = g - \{(\text{normal gravity at 0}) - (\text{normal free-air reduction})\} \\ & = g - (\text{normal gravity at G}), \end{aligned}$$

what has been called the free-air anomaly is not the value to be assigned to the point 0 but instead is to be assigned to the point G. So, strictly speaking, the free-air anomaly should be considered to be a station anomaly rather than an anomaly on the geoid, so far as the normal value of gradient 3086×10^{-9} is used.

References and further reading

Elkins, T. A. 1951. The Second Derivative Method of gravity interpretation. *Geophys.* **16**, 29.

Eötvös, R. v. 1906. Bestimmung der Gradienten der Schwerkraft und Ihrer Niveauflächen mit Hilfe der Drehewage. *Verh. XV Allg. Konf. Intern. Erdmag.*, Budapest **1**, 337.

Evjen, H. M. M. 1936. The place of the vertical gradient in gravity interpretation. *Geophys.* **1**, 127.

Hagiwara, Y. 1966. Three-dimensional distribution of real Bouguer anomalies from gravity values observed at various elevations. *Bull. Earthquake Res. Inst.* **44**, 519.

Henderson, R. G. and L. Cordell 1971. Reduction of unevenly spaced potential field data to a horizontal plane by means of finite harmonic series. *Geophys.* **3**, 856.

Tajima, H. and S. Izutuya 1971. Measurement of the vertical gradient of gravity at Keio Plaza Hotel building (in Japanese). *J. Geod. Soc., Japan* **17**, 187.

Tsuboi, C. 1954. A study of the anomalies in the vertical gradient of gravity with the aid of the Bessel–Fourier Series. *Proc. Jap. Acad.* **30**, 453.

Tsuboi, C. 1955. Anomalies in $\partial g/\partial z$ (in Japanese). *J. Geod. Soc., Japan* **2**, 21.

Tsuboi, C. 1963. Derivation of real Bouguer anomalies from gravity values observed at various elevations (in Japanese). *J. Geod. Soc., Japan* **10**, 94.

Tsuboi, C. 1965. Calculation of Bouguer anomalies with due regard to the anomaly in the vertical gravity gradient. *Proc. Jap. Acad.* **41**, 386.

Tsuboi, C. and M. Kato 1952. The first and second vertical derivatives of gravity. *J. Phys. Earth* **1**, 95.

12 Time variations of gravity

12.1 Periodic and non-periodic variations

The intensity and direction of gravity at a point on the Earth do not remain constant but change with time in small amounts, either periodically or non-periodically.

The periodic variations result from two causes: directly from the effects of the attraction forces of the Moon and the Sun, and indirectly from an elastic deformation of the Earth which is produced by these forces. The direct effect of the Moon or the Sun which is seen at a point on the Earth is produced by the difference of its attractions at that point and at the centre of the Earth. Referring to Figure 12.1, the difference of the forces of attraction at a point P facing the Moon T and at the Earth's centre O is

$$GM_m\left\{\frac{1}{(r-R)^2}-\frac{1}{r^2}\right\} \approx GM_m\frac{2R}{r^3}$$

$$= 2g\left(\frac{R}{r}\right)^3\frac{M_m}{M_e}, \qquad (12.1)$$

assuming that the Earth is perfectly rigid. M_m and M_e are the masses of the Moon and of the Earth respectively. By putting

$$g = G\frac{M_e}{R^2} = 980$$

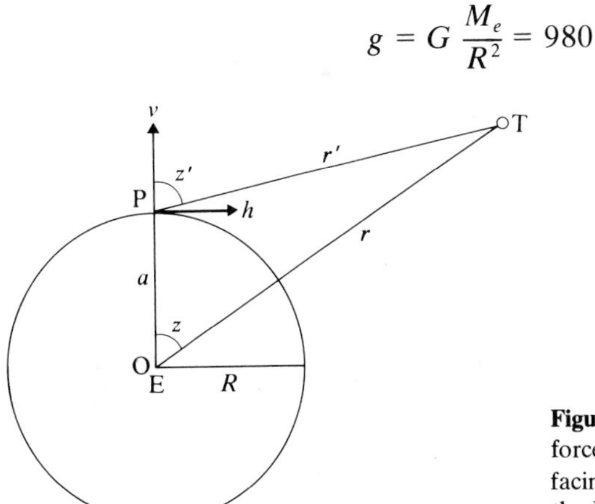

Figure 12.1 Difference of the forces of attraction at a point P facing the Moon (or Sun) T, and at the Earth's centre E.

$$\frac{R}{r} = \frac{1}{60}$$

$$\frac{M_m}{M_e} = 0.012,$$

the lunar variation in gravity at that point is seen to be nearly 1×10^{-4} gal (0.1 mgal) in amplitude. In the case of the Sun, the amplitude is 0.04 mgal by similar calculations.

Figure 12.2 is a record of the variation in gravity due to the Moon and the Sun taken by Y. Tomoda in 1978, using his vibration-string gravity meter. The amplitudes of variation in gravity that are actually observed do not agree with those calculated in the above, where we have made the simple assumption of a perfectly rigid Earth. This is because there are indirect effects that are caused by an elastic deformation of the Earth produced by the forces of the Moon and the Sun. This phenomenon of deformation of the Earth is called the Earth tide, and it will be the subject of a later section.

There are also non-periodic variations in gravity which are produced by several causes other than astronomical. If the level of underground water in a certain area rises as a result of a heavy rainfall, the attraction due to the additional water will increase gravity values over that area. If the void ratio of the soil in the area is 20%, 1 metre rise of the underground-water level will increase gravity value by its attraction,

$$2\pi G \times 100 \times 0.2 \approx 0.01 \text{ mgal} = 10\ \mu\text{gal}.$$

Another instance of the change in gravity is that which is caused by atmospheric pressure variations. A strong atmospheric low pressure is an anomalous deficiency of air mass and the resulting decrease in upward attraction of the air will increase the gravity value. In extreme

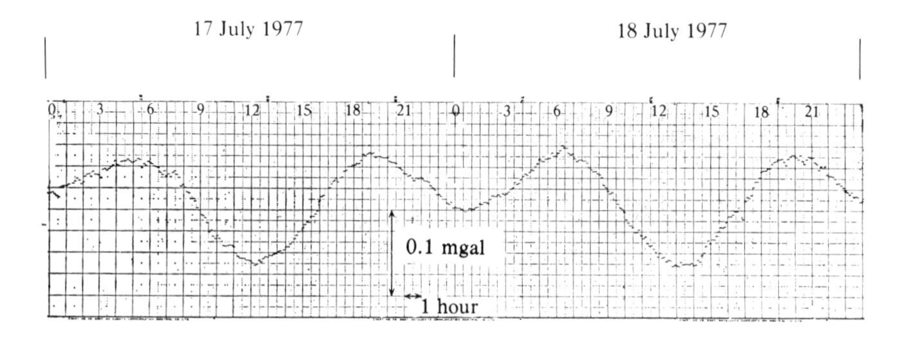

Figure 12.2 Tidal variation in gravity due to the Moon and Sun at Tokyo.

cases, the increase in gravity may amount to several tens of a μgal (which is equal to 10^{-6} cm s^{-2}). It should always be borne in mind that the precise value of gravity measured at a point at one time cannot escape from the effects of such irregular meteorological disturbances.

The changes in gravity which have been enumerated so far are those whose causes are known. There are also changes for which the causes are not yet very clear. There are reports that gravity values change slightly in areas where a large earthquake or a volcanic eruption take place. These changes when established will provide a clue to finding what is taking place underground in association with these geophysical phenomena.

12.2 Tide-generating potentials of the Moon and the Sun

The positions of the Moon and the Sun relative to the Earth, and consequently their attractions, change with time and cause periodic variations in the intensity and direction of gravity at a point on the Earth. The effects of the Moon or the Sun which are observed on an undeformable Earth's surface are the difference of its attractions at that point and at the centre of the Earth. Using the notation shown in Figure 12.3, the changes in gravity are:

$$v = GM\left(\frac{\cos z'}{r'^2} - \frac{\cos z}{r^2}\right) \tag{12.2}$$

in the vertical direction, and

$$h = GM\left(\frac{\sin z'}{r'^2} - \frac{\sin z}{r^2}\right) \tag{12.3}$$

in the horizontal direction, where M is the mass of the Moon or the Sun. z' is called its zenith distance. Given that

$$r'^2 = r^2 - 2ar\cos z + a^2, \tag{12.4}$$

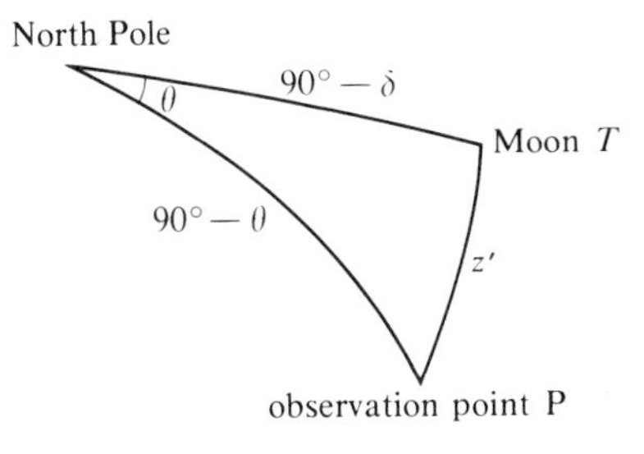

Figure 12.3 Changes in gravity due to the position of the Moon or the Sun, T, and of attractions at an observation point, P, and the centre of the Earth, E. δ = declination; z' = zenith distance; φ = latitude; Θ = hour angle.

$$\sin z' = \left(\frac{r}{r'}\right) \sin z,$$

and also that a is very small compared with r, the above expressions for v and h reduce to

$$v = 3GM \frac{a}{r^3} (\cos^2 z' - \tfrac{1}{3}),$$

$$h = \tfrac{3}{2} GM \frac{a}{r^3} \sin 2z'.$$

Putting

$$g = G \frac{M_e}{R^2}$$

into v and h, we obtain

$$v = 3g \left(\frac{M}{M_e}\right) \frac{aR^2}{r^3} (\cos^2 z' - \tfrac{1}{3}), \tag{12.5}$$

$$h = \frac{3}{2} g \left(\frac{M}{M_e}\right) \frac{aR^2}{r^3} \sin 2z'. \tag{12.6}$$

Both v and h change with z' according to time. $|v|$ is largest when $z' = 0$ or π and its maximum value is

$$|v|_{\max} = 2g \left(\frac{M}{M_e}\right) \frac{aR^2}{r^3}. \tag{12.7}$$

At points on the Earth where z' cannot be 0 or π because of their high latitudes, v becomes a maximum when z' is closest to 0 or π. Here a and R are similar quantities but they differ in that a is the radius vector drawn to the point from the Earth's centre and is considered to be a variable, while R is the mean radius of the Earth which is a constant. In the case of the Moon,

$$M/M_e = 0.0123$$

$$R/r = 1/60.3,$$

therefore

$$|v|_{max} = 2 \times 980 \times 0.0123 \times (10/603)^3 = 0.109 \text{ mgal}.$$

In the case of the Sun,

$$|v|_{max} = 0.041 \text{ mgal}$$

by similar calculations.

The horizontal component h becomes a maximum when $z' = \pm \pi/4$ and the value of the maximum is

$$|h|_{max} = \frac{3}{2} g \left(\frac{M}{M_e}\right) \frac{aR^2}{r^3}, \tag{12.8}$$

which is 0.082 mgal for the Moon and 0.037 mgal for the Sun.

Also the direction of the vertical at a point on the Earth changes due to the attraction of the Moon or the Sun. Its amount θ is h divided by g, so that

$$|\theta|_{max} = \frac{|h|_{max}}{g} = 0.082 \times 10^{-3} \div 980 = 0.017'' \tag{12.9}$$

for the Moon, and is

$$|\theta|_{max} = 0.037 \times 10^{-3} \div 980 = 0.008'' \tag{12.10}$$

for the Sun. We have seen that

$$v = 3g \left(\frac{M}{M_e}\right) \frac{aR^2}{r^3} (\cos^2 z' - \tfrac{1}{3}),$$

$$h = \frac{3}{2} g \left(\frac{M}{M_e}\right) \frac{aR^2}{r^3} \sin 2z'.$$

If the following quantity V is introduced

$$V = \frac{3}{2} g \left(\frac{M}{M_e}\right) \frac{a^2R^2}{r^3} (\cos^2 z' - \tfrac{1}{3}), \tag{12.11}$$

then we see

$$v = \frac{\partial V}{\partial a} \tag{12.12}$$

and

$$h = \frac{\partial V}{a\, \partial z'}. \tag{12.13}$$

Both v and h can be derived by differentiating V in the directions of a and az' respectively. This V is called the tide-generating potential. In the above expression, V is given as a function of z'. If, instead of z', the latitude φ of the observation point, the declination δ of the Moon or the Sun and its hour angle Θ are used, V becomes

$$V = \frac{3}{4} g \left(\frac{M}{M_e}\right) \frac{a^2 R^2}{r^3} \times \{3(\sin^2 \delta - \tfrac{1}{3})(\sin^2 \varphi - \tfrac{1}{3})$$

$$+ \sin 2\delta \sin 2\varphi \cos \Theta + \cos^2 \delta \cos^2 \varphi \cos 2\Theta\}. \tag{12.14}$$

For a given gravity station, the latitude φ is fixed. δ is the declination of the Moon or the Sun and it changes slowly with time (with a period of about 28 days for the Moon and of about one year for the Sun). Θ changes by the combined effects of the rotation of the Earth and the right ascension of the Moon or the Sun. Taking all these into account, the tide-generating potential at a point changes in a very complicated way with time. In order to avoid this, it is customary to decompose the variations of V into a number of what are called component tides, in such a way that each of which changes simple harmonically with time. These component tides are listed in Table 12.1.

Table 12.1 Component tides.

Symbol	*Name*	*Period*	*Relative amplitude*
M_2	principal lunar	12.42 h	0.454
S_2	principal solar	12.00 h	0.212
N_2	larger lunar elliptic	12.66 h	0.088
K_2	lunisolar	11.97 h	0.058
K_1	lunisolar	23.93 h	0.266
O_1	larger lunar	25.82 h	0.189
P_1	larger solar	24.07 h	0.088
M_f	lunar fortnightly	13.66 days	0.078
S_{sa}	solar semi-annual	$\frac{1}{2}$ y	0.037
	nineteen yearly	19 y	0.033

12.3 Tide-generating potential on a deformable Earth

The tide-generating potential V changes with time and it affects various quantities which are related to gravity on the Earth. Actual changes that are observed do not agree in magnitude with the changes that are predicted simply from the change in V for an undeformable Earth. The reason for this is that the Earth is not perfectly rigid and subject to an elastic deformation produced by the forces of the Moon and the Sun.

Let the bold line in Figure 12.4 represent the undeformed shape of the Earth and the lighter line its deformed shape. The gravitational potential at a point P changes not only by the direct effect of the change in V itself but also by the effects produced by the mass difference between the undeformed and deformed Earths as well as by the radial displacement u of the point from P to P′. The whole potential V' at a point is the sum of the

potential due to the undeformed Earth	W
tide-generating potential	V
potential produced by the deformed part of the Earth	ΔW
potential change produced by the radial displacement u	$-u \dfrac{\partial W}{\partial a}$

so that

$$V' = W + V + \Delta W - u \frac{\partial W}{\partial a}. \tag{12.15}$$

ΔW may be regarded to be proportional to V, so that

$$\Delta W = kV$$

u may be regarded to be proportional to V, so that

$$-u \frac{\partial W}{\partial a} = ug = hV.$$

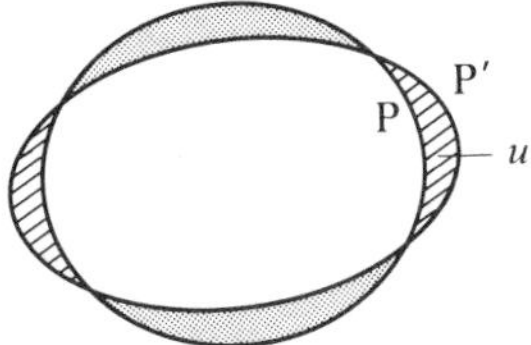

Figure 12.4 Deformation of the Earth by the Earth tide, where the bold line represents the undeformed Earth and the lighter line represents the deformed Earth.

Thus we see finally

$$V' = W + V + \Delta W - u \frac{\partial W}{\partial a}$$

$$= W + V(1 + k - h). \qquad (12.16)$$

The actual variation of gravity potential on a deformable Earth is $(1 + k - h)$ times that calculated on the assumption of a perfectly rigid Earth. k and h are physical constants related to the elastic property of the Earth and were first introduced by Love (1911) and are called **Love's numbers**. If the Earth is made up of an incompressible material of uniform density ρ and rigidity μ, calculations show that

$$k = \frac{3}{2} \bigg/ \left(1 + \frac{19\mu}{2g\rho R}\right),$$

$$h = \frac{5}{2} \bigg/ \left(1 + \frac{19\mu}{2g\rho R}\right). \qquad (12.17)$$

The simplest phenomenon that is produced on the Earth by the Moon and the Sun is the ebb and flow of the sea water. If the Earth were perfectly rigid, the amplitude of variation of the seawater level would be V/g. But since the Earth is not perfectly rigid, the observed amplitude of the sea-level change is $(1 + k - h)V/g$. If the observed amplitude of the sea-level change is 90 cm, while that calculated on the assumption of a rigid Earth is 100 cm, the difference 100 − 90 = 10 cm is a result due to the tidal deformation of the Earth. In this way, the value of the factor $(1 + k - h)$ can be found and from it, the rigidity of the Earth can be estimated. But in actual fact the problem is not that simple. The sea water has its dynamical inertia and also frictional forces are acting between it and the ocean floor. For these reasons, the sea water does not follow the tide-generating potential faithfully. In order to estimate the values of h and k, short-period components of tidal variation are not very useful, but there is a tidal component with a period of 19 years. Lord Kelvin (1824–1907) used this component and compared the amplitudes of the tide-generating potential and of the observed tidal range. From the comparison, he concluded that the average rigidity of the Earth is nearly equal to that of ordinary steel. This is to be remembered as a pioneering study in this field, but his calculations were based on a very simple model of the Earth which is hardly acceptable with our present knowledge about it. So his figure for the Earth's rigidity differs from more recent values.

12.4 Tidal variation of gravity

It was shown that the gravitational potential V' when the Earth tide is taken into consideration is

$$V' = W + V + \Delta W - u\frac{\partial W}{\partial a}. \tag{12.18}$$

Differentiating V' with respect to $-a$, we obtain

$$-\frac{\partial V'}{\partial a} = g + \delta g$$

$$= -\frac{\partial W}{\partial a} - \frac{\partial V}{\partial a} - \frac{\partial(\Delta W)}{\partial a} + u\frac{\partial^2 W}{\partial a^2}.$$

The first term on the right-hand side of the above equation is

$$-\frac{\partial W}{\partial a} = g.$$

ΔW in the third term is due to the mass of the deformed part of the Earth and is expressed generally by

$$\Delta W = 4\pi G\sum_n \frac{A_n}{2n+1}\frac{R^{n+2}}{a^{n+1}}S_n, \tag{12.19}$$

where $S_n(\theta, \lambda)$ is a solution of Laplace's equation expressed in surface spherical harmonics. Since in the deformation of the Earth, the term with $n = 2$ is predominant, we may write

$$\Delta W = 4\pi G\frac{A_2}{5}\frac{R^4}{a^3}S_2, \tag{12.20}$$

so that

$$\frac{\partial(\Delta W)}{\partial a} = -3 \times 4\pi G\frac{A_2}{5}\frac{R^4}{a^4}S_2 = -\frac{3\Delta W}{a}.$$

Since

$$\Delta W = kV,$$

it follows that

$$\frac{\partial(\Delta W)}{\partial a} = -\frac{3}{a}kV.$$

Also since $\partial V/\partial a$ is

$$\frac{\partial V}{\partial a} = \frac{2V}{a},$$

therefore

$$\frac{\partial(\Delta W)}{\partial a} = -\frac{3}{a}kV = -\tfrac{3}{2}k\frac{\partial V}{\partial a}.$$

Finally, the fourth term on the right-hand side of equation (12.18) is

$$u\frac{\partial^2 W}{\partial a^2} = -u\frac{\partial g}{\partial a}.$$

But since we put

$$u = \frac{h}{g}V,$$

$$g = \frac{h}{u}V,$$

$\partial g/\partial a$ reduces to

$$\frac{\partial g}{\partial a} = \frac{h}{u}\frac{\partial V}{\partial a},$$

so that

$$u\frac{\partial^2 W}{\partial a^2} = -u\frac{\partial g}{\partial a} = -h\frac{\partial V}{\partial a}.$$

We have finally

$$g + \delta g = g - (1 - \tfrac{3}{2}k + h)\frac{\partial V}{\partial a}. \tag{12.21}$$

Observed changes in the intensity of gravity are $(1 - \frac{3}{2}k + h)$ times compared with those calculated on the assumption of a perfectly rigid Earth. By comparing the observed and astronomical variations of gravity, the value of the factor $(1 - \frac{3}{2}k + h)$ can be found. The amplitude of variation of g is of the order of 0.1 mgal. In other words, the value of gravity observed at a point is always affected by this term, so that if we wish to find the value of gravity that is free from this effect, we have to apply corrections to observed values. Several numerical tables have been published which are useful for finding these corrections from the hour angles of the Moon and the Sun, the latitudes and longitudes of gravity stations. In these tables, the value of the factor $(1 - \frac{3}{2}k + h)$ is taken to be 1.2.

12.5 Tilting of the ground

The direction of the surface of the ground changes with time for various reasons. The variation is of the order of several seconds of angle in amplitude and is easily observed by instruments called tiltmeters. The main part of the variation is due to the solar radiation which causes thermal deformation of the ground surface. But even in a deep vault which is free from the effect of solar radiation, the variation is still observed.

A tiltmeter measures changes in the angle between the directions of the vertical at a point and of the plane on which it is installed. If the direction of the vertical does not change, the instrument will record the geometrical tilt of the ground. But on the Earth, the direction of the vertical itself is not fixed and changes according to the attractions of the Moon and the Sun. In addition to this, the direction of the plane on which a tiltmeter is set up changes according to the Earth tide. What is observed by a tiltmeter is a combined effect of the change in the direction of the vertical due to the tide-generating forces and the elastic deformation of the Earth. According to calculations, the observed tilt is $(1 + k - h)$ times the astronomical change in the direction of the vertical. By comparing the observed tilt with the astronomical variation of the vertical, the value of this factor can be found.

If tilt observations are made near a sea coast, other complexities come into the phenomena. The attraction of the seawater mass has an influence on the direction of the vertical. Not only that, the seawater mass will act as a load applied to the ground and will make it deform.

12.6 Latitude variation

The latitude of a point on the Earth is affected by the attractions of the Moon and the Sun. The amplitude of the variation is of the order of 0.01 second of arc. There are two reasons for this variation. One is the effect of the change in the direction of the vertical, to which astronomical observations are referred, and the other is the horizontal displacement of the point in the meridional direction due to the elastic deformation of the Earth. The observed variation is a combined effect of the two and is $(1 + k - l)$ times the variation in the vertical direction on a rigid Earth. The number l is a constant which was first introduced by Shida (1912) and is called **Shida's number**. According to calculations,

$$l = \frac{3}{4} \Big/ \left(1 + \frac{19}{2} \frac{\mu}{\rho g R}\right), \tag{12.22}$$

for an Earth of a uniform density ρ and rigidity μ.

12.7 Other periodic variations related to Earth tides

There are several other observable quantities which are related to the elastic properties of the Earth. They are not directly related to the variation of gravity but give us information about the elasticity of the Earth. These are the phenomena of the nutation of the Earth, horizontal strain and volume change of the Earth.

The rotation axis of the Earth makes wobbling motions which are called the **free nutation**. As a result, the latitude of a point is observed to change, or what is the same thing, the rotation axis of the Earth is moving with respect to the Earth. The curve in Figure 12.5 shows the migration of the rotation axis with respect to the Earth, observed from 1971 to 1975. During these five years, the position of the rotation axis made four loops so that one loop is completed in about 450 days. Data from longer observation give 430 days. This length is called **Chandler's period** T_C (S. H. Chandler, 1846–1913) after the name of the astronomer who studied this problem. Before this, L. Euler (1707–1783) had calculated the nutation period of a rigid Earth; the period obtained by him was 305 days. This is called **Euler's period** T_E and is much shorter than T_C. The disagreement is due to the effect of the elastic deformation of the Earth by the Earth tide. The ratio T_E/T_C is

$$\frac{T_E}{T_C} = 1 - k \Big/ \left(\frac{2g\varepsilon}{R\omega^2} - 1\right), \tag{12.23}$$

using one of Love's numbers k. Here ε is the ellipticity of the Earth, R its equatorial radius and ω its angular velocity of rotation. By comparing T_E and T_C, the value of k can be estimated.

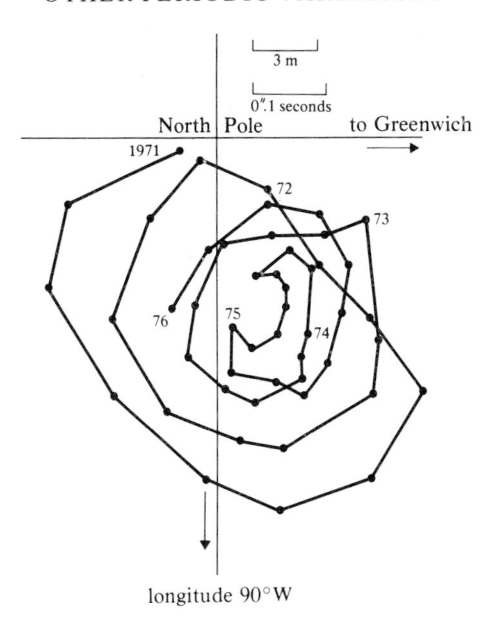

Figure 12.5 Curve showing free nutation, a wobbling of the rotational axis relative to the Earth, observed from 1971 to 1976.

The distance between two points on the Earth's surface changes according to the elastic deformation of the Earth produced by the Earth tide. Horizontal strains at a point are:

$$e_\varphi = \frac{l}{Rg}\frac{\partial^2 V}{\partial \varphi^2}$$

in the direction of latitude and

$$e_\lambda = \frac{l}{Rg\cos^2\varphi}\frac{\partial^2 V}{\partial \lambda^2}$$

in the direction of longitude, where l is Shida's number; e is of the order of 10^{-8} and can be observed by an extensometer or strain meter. From observations of horizontal strains, the value of l can be estimated.

The volume of a certain portion of the Earth also changes according to the Earth tide. According to calculations, the dilatation of volume is

$$\Delta = \frac{4V}{gR}(h - \tfrac{3}{2}l),$$

and the value of the factor $(h - \frac{3}{2}l)$ can be found by comparing observed Δ and calculated V. Dilatation of volume near the surface of the ground

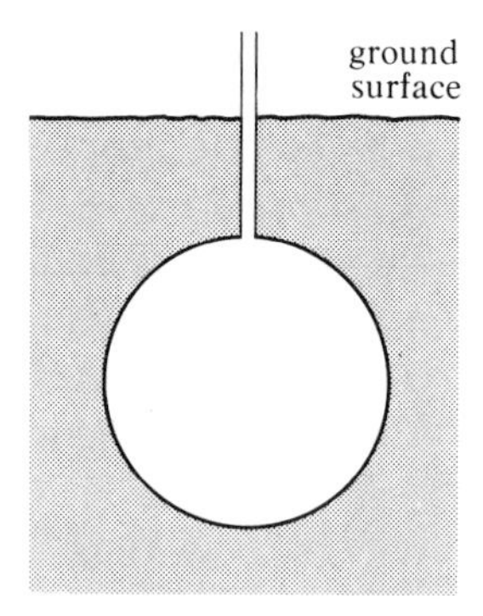

Figure 12.6 Apparatus for measurement of the dilatation of volume near the surface of the ground.

can be measured by an arrangement shown in Figure 12.6. A big container is embedded in the ground and filled with a suitable liquid. A thin tube is connected to the container and the liquid is led into it. Up and down motions of the liquid meniscus record the volume change of the ground in which this instrument is installed.

12.8 Earth tides and the structure of the Earth

Summarising what has been discussed, we can discover:

$1 + k - h$	from tidal height variations;
$1 - \frac{3}{2}k + h$	from gravity variations;
k	from nutation of the Earth;
$1 + k - h$	from tiltmeter observations;
$1 + k - l$	from latitude variations;
l	from horizontal strain;
$h - \frac{3}{2}l$	from volume change.

Some of these can be determined accurately, while some other values are not so precise. Takeuchi (1950) carefully investigated all the available observational data and concluded that the most probable values of h, k and l are:

$$h = 0.61 \qquad k = 0.29 \qquad l = 0.08$$

As we have seen, the values of h, k and l for an Earth of uniform density ρ and rigidity μ are:

$$h = \frac{5}{2}\frac{1}{p}, \qquad k = \frac{3}{2}\frac{1}{p}, \qquad l = \frac{3}{4}\frac{1}{p},$$

where

$$p = 1 + \frac{19}{2}\frac{\mu}{\rho g R}.$$

These simple relations are not satisfied by Takeuchi's values. This is because the above relations among h, k and l have been derived on the assumption of a uniform density and rigidity of the Earth. The actual Earth is far from such an idealized model. K. E. Bullen (1906–1976), from the velocity distribution of earthquake waves within the Earth, obtained density and rigidity distributions at various depths. Takeuchi, in 1950, elaborately calculated the values of h, k and l according to Bullen's model. The final values obtained by Takeuchi are

$$h = 0.59\text{–}0.61, \qquad k = 0.28\text{–}0.29, \qquad l = 0.07\text{–}0.08$$

showing very good agreements with the observed values.

Earthquake S waves are not propagated through the core of the Earth and in this sense, the core of the Earth may be said to act like a liquid. Takeuchi assumed various values for the rigidity of the core and found that, if the rigidity is $0 \sim 10^{10}$, the calculated values of h and k do not contradict the observed values.

12.9 Non-periodic variations of gravity

Gravity variations produced by external forces of the Moon and the Sun are periodic in nature and can be explained by the theory of dynamics, as explained in the preceding sections. Other than these, there are indications that gravity values and also Love's numbers, h and k, in an area change with time in connection with non-systematic geophysical phenomena.

The first instance is related to fluctuations in the level of underground water. Since 1963, H. Tazima and others have repeated measurements of the gravity difference between Tokyo (University of Tokyo) and Kakioka (Magnetic Observatory), about 80 km north-east of Tokyo, by means of gravimeters. Tokyo is situated on young sedimentary layers while Kakioka lies on a solid granite mass. They have found that $(g_T - g_K)$ has been gradually changing, as shown in Figure 12.7, with an annual rate of -11 μgal per year. There is a deep-water well on the campus of the university and the level of underground water in it has been continuously measured. As is seen in Figure 12.7, the water level has been lowering with an average rate of 2 m/year. Tazima showed that the apparent relative decrease of gravity at Tokyo with respect to

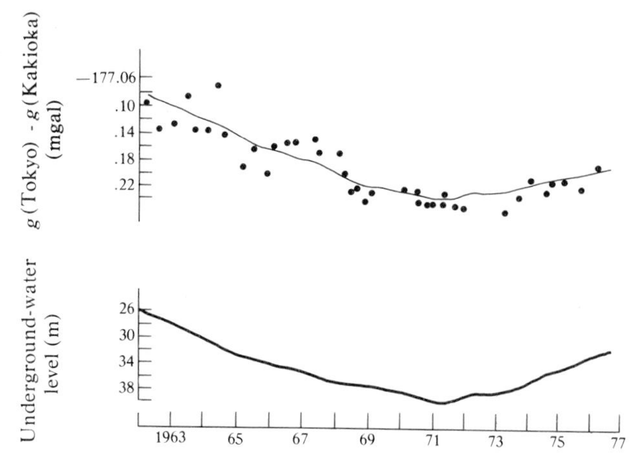

Figure 12.7 Decrease in gravity at Tokyo with respect to that at Kakioka compared with the decrease in the underground-water level at Tokyo between the years 1963 and 1976.

Kakioka can be accounted for by the decrease of attraction due to the lowering of the underground mass of water. He took the void ratio of the soil in Tokyo to be 15%. The annual decrease of the mass is then

$$200 \times 0.15 = 30 \text{ cm}$$

and the resulting decrease in gravity will be

$$2\pi G \times 30 = 11 \ \mu\text{gal}$$

which agrees well with the observed changes.

The second instance of gravity change is related to the change in height of gravity point. In an earthquake-prone country such as Japan, where the ground is not geologically stable, its heights vary in amounts detectable by precise levelling, especially in connection with earthquakes. Figure 12.8 shows the variation in the heights of the ground (1969–1976) in a volcanic area of Idu which is about 100 km south-west of Tokyo. During 1974–5, a large number of small earthquakes occurred in this area and along with the precise levelling, gravity values were measured also. At four points in this area, changes in height and gravity were both measured and the points plotted in Figure 12.9 show the relationship existing between them. As is seen in the figure, upheaval of the ground is associated with a decrease in gravity, the slope being $-3 \ \mu\text{gal cm}^{-1}$. This is equal to the free-air gradient -1 mgal per 3 m and this shows that the change in gravity is caused by a vertical dilatational movement of the ground without any change in the mass of the ground as shown in Figure 12.10.

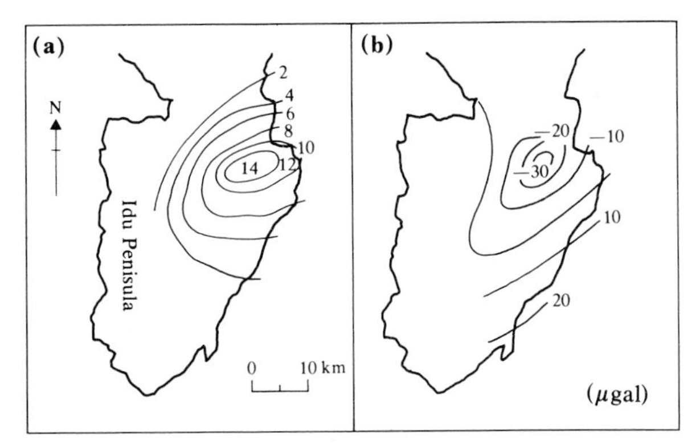

Figure 12.8 Changes in (a) elevation (cm) between 1969 and 1976, and (b) changes in gravity (μgal) between 1974 and 1976, on Idu Peninsula on the Pacific coast of the main island of Japan.

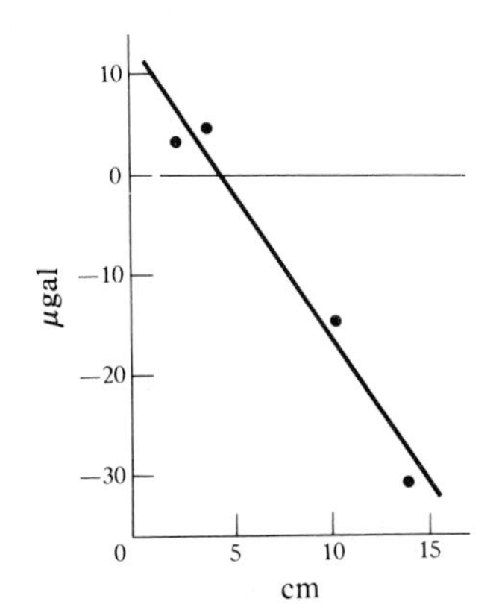

Figure 12.9 Relationship between the change in gravity and the change in elevation on Idu Peninsula.

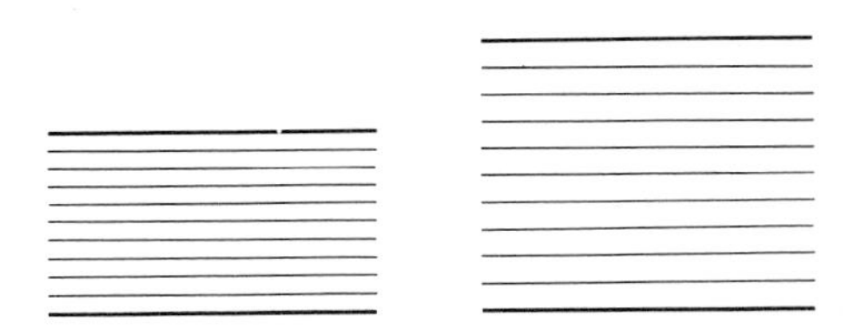

Figure 12.10 Free-air type of vertical elevation of the ground without a change in the mass of the ground.

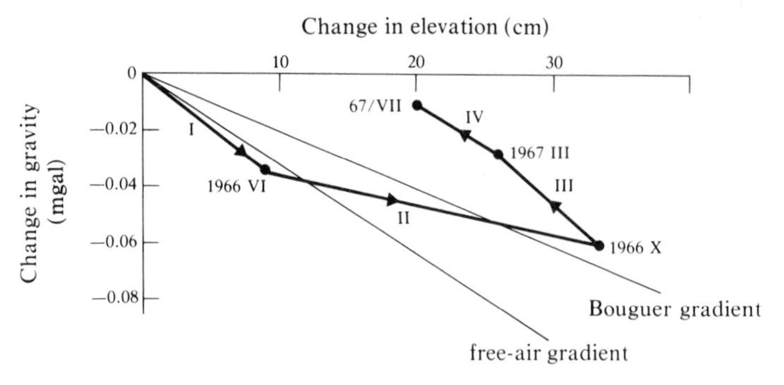

Figure 12.11 Relationship between the changes in gravity and the changes in elevation at Matsushiro, observed between 1966 and 1967.

Another instance of a gravity change was observed in Matsushiro, 150 km north-west of Tokyo, when a large number of small earthquakes occurred in 1965–7, with a maximum number as large as 500 per day. In this case also, both gravity values and heights were measured. The points plotted in Figure 12.11 show the relationship between these changes. The slopes for the periods I, III and IV are close to the free-air gradient but the slope for the period II is remarkably smaller than this. Instead of a mere dilatation of a small portion of the ground, an uplift of the ground of wider extent can be imagined. In this case, the change in gravity is the sum of the effects of gravity increase due to the thickening of the ground and the free-air decrease of gravity due to the increase in height. This change may be called a Bouguer type. Decrease of gravity for the latter reason is smaller than the free-air type for the same amount of uplift. The gravity change at Matsushiro in the second period II is still much smaller than this. In order to explain the low magnitude of the gravity decrease, an intrusion of some liquid, probably of water, into this area must be assumed as an additional mass.

Finally, an instance in which Love's numbers changed will be described. Y. Hagiwara made continuous observations of the tidal variation of gravity at a point on the Pacific coast of Honshu (the main island of Japan), and from these he calculated the constant $(1 - \frac{3}{2}k + h)$ for every day. He found that the value was 1.2 on average, but in May 1974, the value showed an anomalous change as shown in Figure 12.13,

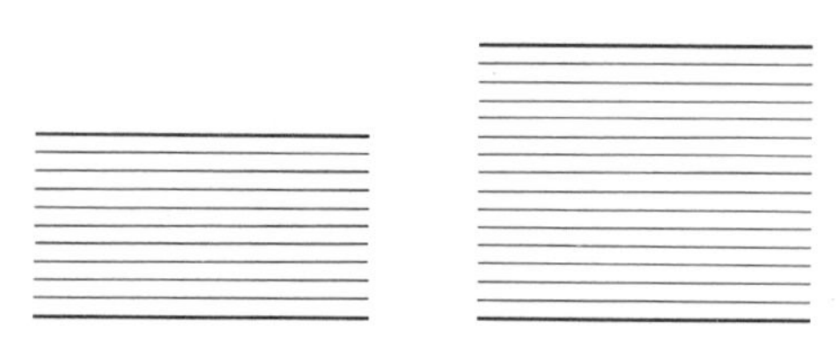

Figure 12.12 Bouguer type of vertical elevation of the ground.

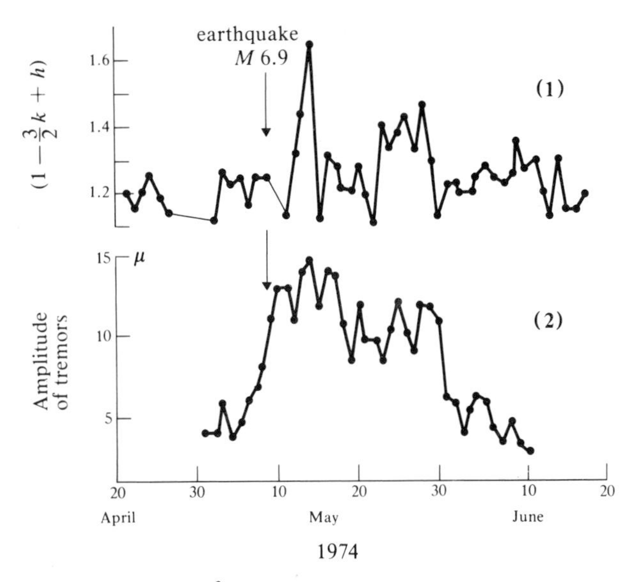

Figure 12.13 Changes in $(1 - \frac{3}{2}k + h)$ at Idu and the amplitude variation of volcanic tremors at volcano Ooshima, 1974.

reaching 1.61. On 9 May, 1974, an earthquake of $M = 6.9$ took place and after a few weeks, the value came to its original value. It is note-worthy that at this time, amplitudes of volcanic tremors on Ooshima volcanic island, not far from the site, also showed an anomalous increase (as shown in Fig. 12.13).

References and further reading

Darwin, G. H. 1907. *Attempted evaluation of the rigidity of the Earth based on tides of long periods*. Sci. Papers, I, Cambridge.

Love, A. E. H. 1911. *Some problems of geodynamics.* Cambridge: Cambridge University Press.

Shida, T. and M. Matuyama 1912. *Change of plumb line referred to the axis of the Earth as found from the result of international latitude observations*. Mem. Coll. Sci. Eng., Kyoto Imp. Univ., IV, no. 1.

Takeuchi, H. 1950. On the Earth tide in the compressible Earth of varying density and elasticity. *Trans Am. Geophys. Union* **31**, 651.

Tsuboi, C. 1929. Observation of the time variation of the second space derivatives of the gravitational potential. *Bull. Earthquake Res. Inst.* **7**, 457.

13 Isostasy

13.1 Outline of isostasy

Isostasy is a condition of statical equilibrium of the Earth's crust which was disclosed by means of observations related to gravity. According to this idea of isostasy, the Earth's crust is 'floating' on a denser substratum as shown in Figure 13.1, and the weights of irregular topographies, such as mountains and oceans, are supported by hydrostatic buoyancy working from the substratum to the crust at its bottom.

When Bouguer anomalies $\Delta g_0''$, observed at thousands of points on the Earth, are plotted against their heights h (negative in the case of the sea), the points lie on two roughly straight lines as shown in Figure 13.2.

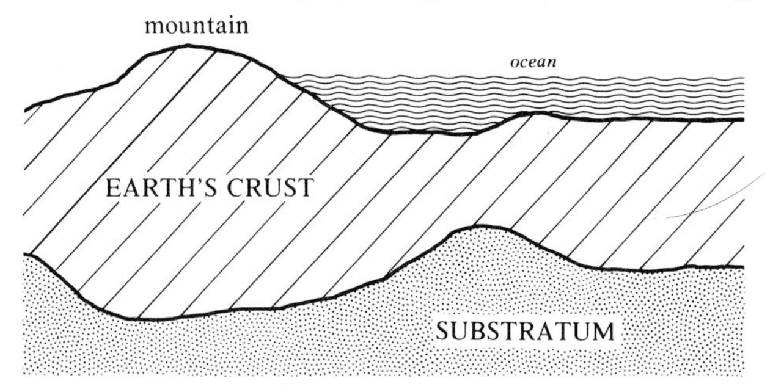

Figure 13.1 Schematic picture of isostasy.

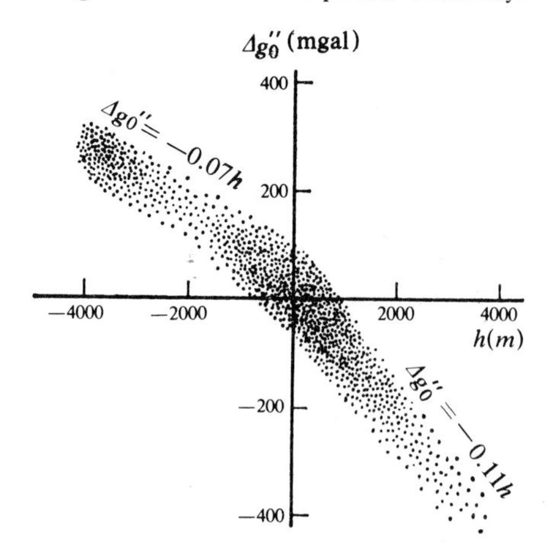

Figure 13.2 Relationship between the Bouguer anomalies and their heights h.

Bouguer anomalies at land points ($h > 0$) are mostly negative and the higher the points, the more negative are the anomalies. In contrast to this, Bouguer anomalies at oceanic points ($h < 0$) are mostly positive, and the deeper the sea under the points, the larger are the positive anomalies. The linear relation for land points can be expressed approximately in mgal by

$$\Delta g_0'' = -0.11h, \qquad h > 0$$

where h is measured in metres and for oceanic points by

$$\Delta g_0'' = -0.07h, \qquad h < 0.$$

For a mountain point $h = 1000$ m, $\Delta g_0''$ is -110 mgal, for instance. This shows that there must be a mass deficiency m under the mountain which produces a gravitational attraction of -110 mgal. To find m, the Bouguer anomaly of -110 mgal is equated to $2\pi G$ times m:

$$-0.110 = 2\pi Gm$$

and from this, m is found to be

$$m = -2.6 \times 10^5 \text{ g}.$$

On the other hand, the mountain mass with a density 2.67 and a thickness of 1000 m is 2.67×10^5 g. The mass of the mountain is seen to be just equal to the mass deficiency beneath the mountain. This coincidence can be interpreted as indicating that the Earth's crust with a density ρ is floating on a substratum with a higher density ρ'. The huge mass of a mountain is supported by the buoyancy acting on that part of the Earth's crust which bulges downward into the substratum, in just the same way as an iceberg keeps its equilibrium floating on the sea water. Referring to Figure 13.3, the condition for hydrostatic buoyancy equilibrium of the ligher Earth's crust of a density ρ floating on the denser substratum of a density ρ' is

$$\rho h = (\rho' - \rho)d, \tag{13.1}$$

or

$$d = \frac{\rho}{\rho' - \rho}h.$$

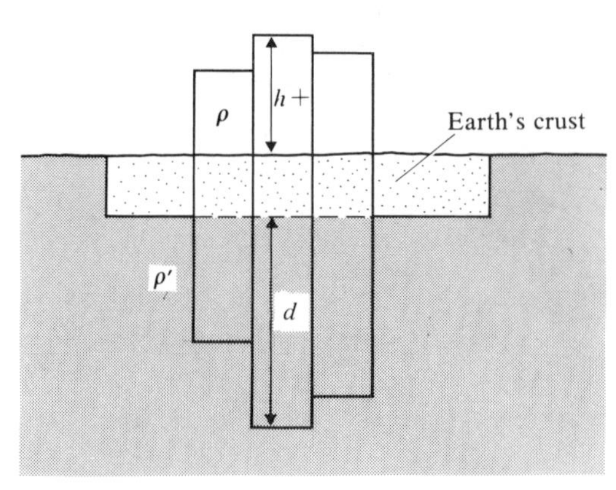

Figure 13.3 Hydrostatic buoyancy equilibrium of the lighter Earth's crust floating on the denser substratum.

If $\rho = 2.67$ and $\rho' = 3.00$, then

$$d = \frac{2.67}{3.00 - 2.67} \approx 8h.$$

This means that the interface between the bottom of the Earth's crust and the substratum has a much larger amplitude of undulation than the surface topography.

From Figure 13.2, $\Delta g_0''$ at an oceanic point $h = -1000$ m, for instance, is +70 mgal. There must be a (relatively) positive mass m under the point by which an attraction of +70 mgal is produced. From

$$0.070 = 2\pi G m$$

m is found to be

$$m = 1.7 \times 10^5 \text{g}.$$

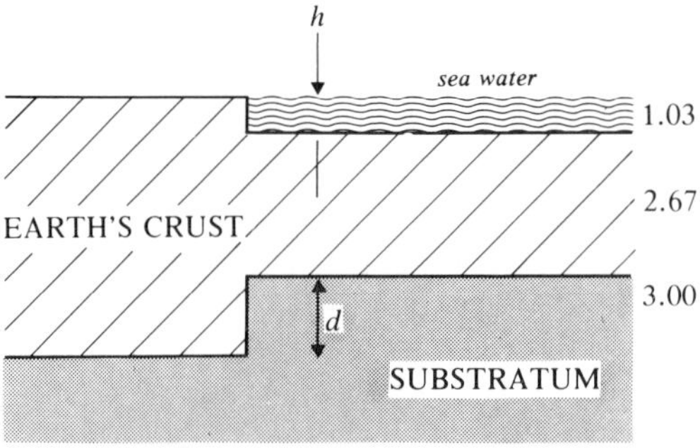

Figure 13.4 Hydrostatic equilibrium of the oceanic crust.

Referring to Figure 13.4, there is the sea water which is 1.03 in density under an oceanic point, instead of the crustal material of $\rho = 2.67$. This density difference between the sea water and the crustal material represents a relatively negative mass

$$(1.03 - 2.67) \times 10^5 = -1.6 \times 10^5 \text{ g}$$

for 1000 m. This is seen to be just equal to m with the algebraic sign reversed. From Figure 13.4, the condition for hydrostatic equilibrium of the oceanic Earth's crust is

$$(2.67 - 1.03) \times h = (3.00 - 2.67) \times d$$

or

$$d = \frac{2.67 - 1.03}{3.00 - 2.67} = \frac{1.64}{0.33} \approx 5h.$$

The interface between the bottom of the oceanic crust and the underlying substratum has an amplitude of undulation five times as large as the sea bottom topographies. This undulation is not so large as in the case of the continental crust.

If the average depth of the world oceans is taken to be 4 km, the oceanic crust must generally be thinner by $4 \times 5 = 20$ km, as compared with the land crust. All that has been seen above indicates that where there is a positive mass superimposed, there is a compensating negative mass under it and where there is a negative mass like an ocean, there is a compensating positive mass under it. It is clear that they are in hydrostatic balance. This condition is called **isostasy**. Of course, isostasy does not hold perfectly everywhere over the Earth, but there can be no doubt that this is the condition most widely seen in the structure of the upper part of the Earth.

13.2 Historical development of the idea of isostasy

Historically, the idea of isostasy did not originate in connection with observations of gravity values in the way explained above. It originated from observations of the deflections of the vertical in India measured during the 1850s, although there was no such term as 'isostasy' at that time. It was in 1889 that the term was first introduced by C. E. Dutton.

In India at that time, geodetic measurements were actively being made by British surveyors. One of the purposes of the geodetic readings was to see how the deflection of the vertical in that region varies along a meridian $\lambda = 77.7°E$. The results of observations for three points are given in Table 13.1.

Table 13.1 Northward deflections of the vertical in India.

Point of observation	*Latitude*	*Relative value (obs.)*	*Calculated value*	*Relative value (calc.)*
Kaliana	39° 31′ N	3.367″	27.853″	20.944″
Kalianpur	24° 7′ N	−1.869″	11.968″	5.059″
Damargida	18° 3′ N	–	6.909″	–

Andrew Waugh, then the director of the Geodetic Survey of India, thought that the deflections of the vertical at these points were caused by the northward attraction of the Himalayan mountains. He asked Archdeacon J. H. Pratt of Calcutta to calculate the amounts of the deflection of the vertical expected at the three points on this supposition. As is seen from Figure 13.5, the angle θ of deflection of the vertical is

$$\theta = \frac{\text{horizontal attraction of the Himalayas}}{g}.$$

The results of Pratt's calculations (Pratt 1855) are given in Table 13.1. A remarkable point to note is that the observed values of the deflection in the table are much smaller than the calculated values. There is no doubt that the Himalayan mountains exert a horizontal attraction to each of the three points. Then, in order to explain the smallness of the observed deflections compared with those calculated, some relatively negative mass must be assumed to exist under the mountains which will make the apparent total horizontal attractions smaller.

G. B. Airy (1855), then the Astronomer Royal, hearing this, suggested a model as shown in Figure 13.6 to explain this disagreement of the observed and calculated deflections of the vertical. He considered that the deeper part of the Earth is 'lava' and that the 'crust' which covers it is floating on it. Where the outer surface of the crust projects upward as mountains, the bottom surface of the crust was thought to bulge downwards into the lava as shown in Figure 13.6. In this part, the lighter crust replaces the heavier lava, forming a relatively negative mass there and thus decreases the apparent horizontal attraction of the mountains. In Airy's model, the density of the crust was assumed to be the same everywhere.

Several years after Airy, Pratt (1859) proposed another model of the Earth's crust to explain the smallness of the apparent attraction of mountains. Pratt thought that the density of the crust is not uniform but is smaller under high mountains and is larger under low lands and that,

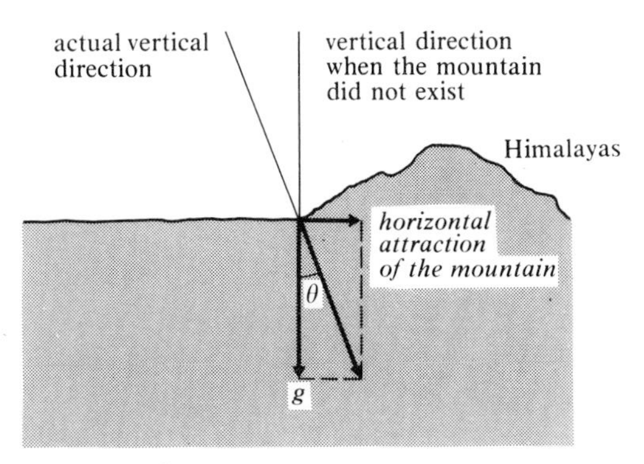

Figure 13.5 Horizontal attraction of the Himalayas causing a deflection of the vertical.

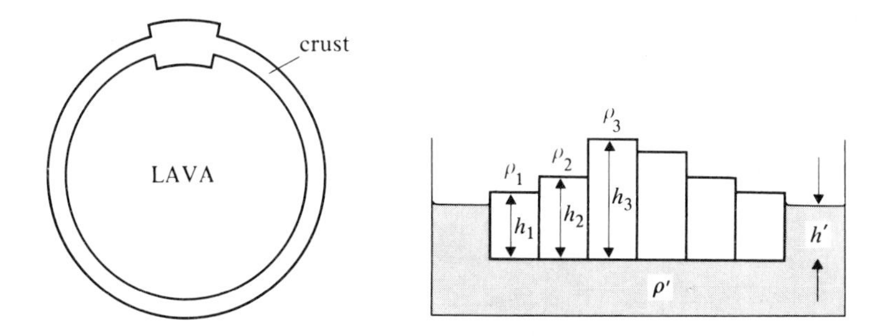

Figure 13.6 Airy's model of isostasy.

Figure 13.7 Pratt's model of isostasy.

at a certain depth from the Earth's surface, the weights of topographical masses become the same irrespective of their heights. Pratt tentatively considered that this depth was about 200 km. In this model also, a relatively negative mass exists under high mountains because of smaller densities, and the disagreement of the observed and calculated values of deflection of the vertical can be explained accordingly. The level at which the pressures under high and low topographies become equal was called the plane of compensation, and its depth was called the depth of compensation.

Referring to Figure 13.7, the conditions for equilibrium of various topographies are

$$\rho_1 h_1 = \rho_2 h_2 = \ldots .$$

Therefore

$$\frac{\rho_2 - \rho_1}{\rho} = -\frac{h_2 - h_1}{h}. \tag{13.2}$$

If $h_2 - h_1 = 1000$ m, and $h_1 \approx h_2 \approx 200$ km (Pratt's value), then

$$\frac{\rho_2 - \rho_1}{\rho} = 0.005.$$

A difference of 1 km in height corresponds to a difference of 0.014 in density provided $\rho_1 = \rho_2 \approx \rho \approx 2.7$.

There are thus two models proposed for explaining the isostatic condition of mountains. Airy's model of the crust may be simulated by wooden blocks with the same density and various lengths floating on water. Pratt's model is similar to the equilibrium of different metallic blocks, whose lengths are inversely proportional to their densities, which are floated on a mercury surface. In Airy's model, the bottom surface of the Earth's crust has an undulation, while in Pratt's model, it is one plane at a constant depth. These models are shown in Figure 13.8.

According to recent studies of the crustal structure of the Earth from the propagation of earthquake waves, Airy's model is believed to be closer to the truth, as will be explained later. Airy's 'lava' is a deformable plastic material and it corresponds to the mantle as we call it now.

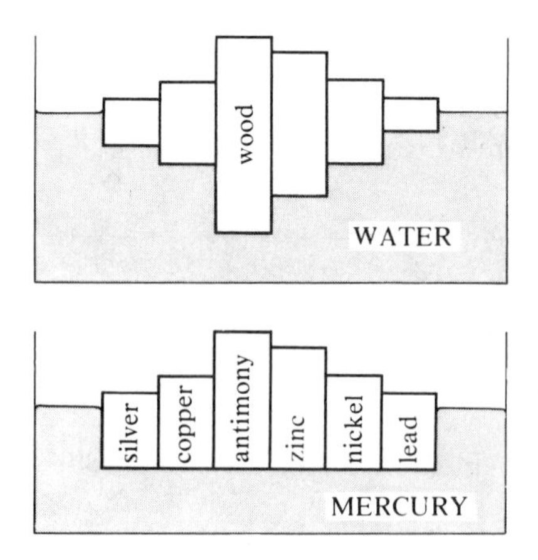

Figure 13.8 Two models proposed for explaining the isostatic condition of mountains: (a) wooden blocks of various lengths, but equal density, floating on water and (b) metal blocks of various densities floating on mercury.

13.3 Investigations made by Hayford, Bowie and Heiskanen

After about 50 years since its discovery, the problem of isostasy began to be intensively studied at the US Coast and Geodetic Survey. Hayford (1909) investigated deflections of the vertical at about 250 points in the

United States, basing his calculations on Pratt's isostasy. He assumed various values for the depth of compensation and compared the observed deflections with the calculations for these depths and tried to find the depth for which $\Sigma(\text{obs.} - \text{calc.})^2$ is a minimum. He found that the sum is a minimum for a depth of 120 km.

After this, W. Bowie, either in co-operation with Hayford (1912) or alone (Bowie 1917), investigated gravity data in the United States. He assumed Pratt's model also and tried to find the depth of compensation for which $\Sigma(\text{obs.} - \text{calc.})^2$ is a minimum. His results also gave a value of 120 km as the most probable depth of compensation, as the curve in Figure 13.9 shows.

In these studies, Bowie not only determined the most probable depth of compensation, but also made one important further step for understanding the mechanism of isostasy. The problem he attacked was whether the isostatic condition holds for every individual small topography (local isostasy) or whether it holds for the average topographical height taken within a certain limiting distance (regional isostasy). Considering that the Earth's crust has some mechanical strength of its own, it is more reasonable to suspect that isostasy holds for the average topographical heights taken within a distance of R rather than to assume it to hold for every small local topography. Bowie gave the name 'regionality' to this limiting extent and his results showed that this R in the USA was a little larger than 100 km. Larger topographies extending more than this are supported by isostasy and topographies smaller than this are supported by the mechanical strength of the Earth's crust.

W. Heiskanen of Finland has contributed greatly to the study of isostasy. Since 1925, he has made extensive investigations of the existing

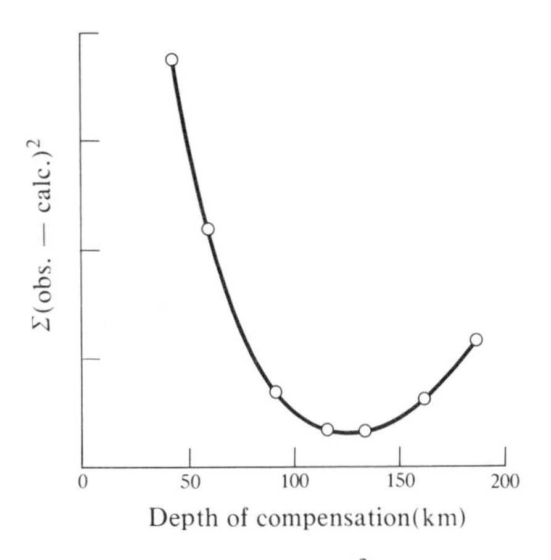

Figure 13.9 Relationship between $\Sigma(\text{obs.} - \text{calc.})^2$ of gravity values in the USA and the depth of compensation.

gravity data over the world, by assuming both Airy's model and Pratt's model, with various values of the thickness of the crust, the depth of compensation, the density distribution within the crust and also of regionality. From these studies, he noticed that Airy's model with a thickness of the crust d gives almost the same results as Pratt's model with a depth of compensation $2d$. It was on an empirical basis that he found this, but this is only natural from a mathematical point of view. It was already proved before that if there is a plane mass distribution at one depth d which produces a certain gravity distribution on the Earth's surface, the same gravity distribution is produced if that mass is uniformly distributed from the surface to a depth $2d$. Airy's model and Pratt's model can therefore not be distinguished from gravity data. Airy's model with a crustal thickness of 60 km, for instance, is just as good as Pratt's model with a depth of compensation of 120 km as far as isostasy is concerned.

13.4 Thickness of the Earth's crust

In subsequent discussion, Airy's model will be adopted throughout. If, for some reason, Pratt's model is preferred to Airy's, conversion between the two models can easily be made by taking into consideration a factor of 2.

In their elaborate studies on the isostatic conditions, Hayford, Bowie and Heiskanen assumed various values for the thickness of the Earth's crust and tried to see which thickness would make $\Sigma(\text{obs.} - \text{calc.})^2$ a minimum. A much simpler and more direct estimation of the thickness is possible by using the Fourier series.

Suppose a surface topography is expressed in a Fourier series

$$H(x) = \sum_m H_m \begin{matrix}\cos\\ \sin\end{matrix} mx \tag{13.3}$$

in a one-dimensional case. According to Airy's model, there is a mass distribution $-\rho$ times $H(x)$ along the bottom surface of the Earth's crust of an average thickness d. The Bouguer-anomaly distribution on the Earth's surface which is produced by this mass distribution is then

$$\Delta g_0''(x) = -2\pi G\rho \sum_m H_m \begin{matrix}\cos\\ \sin\end{matrix} mx \exp(-md). \tag{13.4}$$

On the other hand, let the observed Bouguer-anomaly distribution be expressed by

$$\Delta g_0''(x) = \sum_m B_m \begin{matrix}\cos \\ \sin\end{matrix}\, mx. \tag{13.5}$$

According to Airy's isostasy, the above two expressions for $\Delta g_0''$ are equal, so that

$$-2\pi G\rho H_m \exp(-md) = B_m$$

for each order m. From this, d can be found directly by a relation

$$d = -\frac{1}{m}\log_e\left(-\frac{B_m}{2\pi G\rho H_m}\right). \tag{13.6}$$

There is a minus sign on the right-hand side of equation (13.6). In order that a positive value of d is obtained from this equation,

$$\log_e\left(-\frac{B_m}{2\pi G\rho H_m}\right)$$

should itself be negative and this requires that

$$|2\pi G\rho H_m|$$

should be larger than

$$|B_m|.$$

Also there is a minus sign in

$$\left(-\frac{B_m}{2\pi G\rho H_m}\right).$$

Since

$$\left(-\frac{B_m}{2\pi G\rho H_m}\right)$$

should be positive because its logarithm is to be taken, B_m and the corresponding H_m should be opposite in algebraic sign. There are thus two necessary conditions which a pair of B_m and the corresponding H_m

should satisfy in order that a positive d will result. The pairs of H_m and B_m that satisfy these two conditions will hereafter be called rational pairs and the thickness d of the Airy crust can be calculated using the rational pairs.

This method can easily be extended to two-dimensional cases by using a double Fourier series and the result for these cases is

$$d = -\frac{1}{\sqrt{(m^2+n^2)}}\log_e\left(-\frac{B_{mn}}{2\pi G\rho H_{mn}}\right). \tag{13.7}$$

13.5 Isostasy in North America

The Fourier series method which was mentioned above was applied for studying the isostatic condition in the North American continent by the present author in 1939.

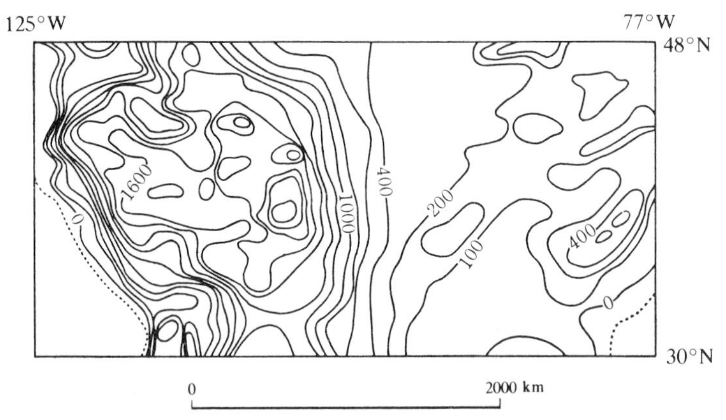

Figure 13.10 Distribution of topographical heights (m) in the USA.

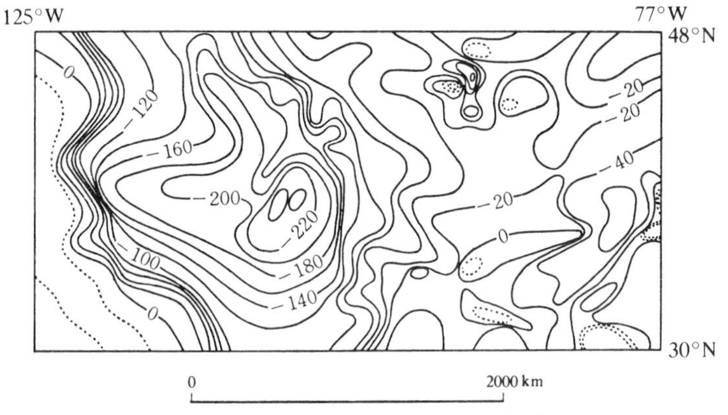

Figure 13.11 Distribution of Bouguer anomalies (gal) in the USA.

A rectangular area 2000 km × 4000 km extending $\varphi = 30°–40°N$, $\lambda = 125°–77°W$ was taken. The topography and Bouguer anomaly distributions in the area are roughly shown in Figures 13.10 and 13.11. As is clearly seen in these figures, $\Delta g_0''$ is negative in high mountainous areas, indicating that isostatic equilibrium holds in these regions, if not perfectly. The distributions of $\Delta g_0''$ and h were expanded into a double Fourier series such as

$$\Delta g_0''(xy) = \sum_m \sum_n B_{mn} \begin{matrix}\cos\\ \sin\end{matrix} mx \begin{matrix}\cos\\ \sin\end{matrix} ny, \qquad (13.8)$$

$$h(xy) = \sum_m \sum_n H_{mn} \begin{matrix}\cos\\ \sin\end{matrix} mx \begin{matrix}\cos\\ \sin\end{matrix} ny, \qquad (13.9)$$

by taking m and n from 0 to 18. There are 18 × 18 × 4 = 1296 pairs of coefficients H_{mn} and B_{mn}, but not all of them are rational pairs and useful for determining the thickness of the Airy crust, because there are pairs that do not satisfy the two necessary conditions that:

(a) B_{mn} and H_{mn} should be opposite in algebraic sign;
(b) $|2\pi G\rho H_{mn}| > |B_{mn}|$.

Among 1296 pairs of H_{mn} and B_{mn}, 712 pairs satisfy condition (a). The percentage of these rational pairs over the total pairs which have the same order depends on the order itself. In this particular case of North America, it must be remembered that the order of a Fourier component should be taken as being $\sqrt{(m^2 + 4n^2)}$ instead of $\sqrt{(m^2 + n^2)}$ because the actual length which corresponds to 2π in the north–south direction is a half of that in the east–west direction. Because of this, n in the north–south direction should be multiplied by two. The dependence of the percentage of the rational pairs on their orders is shown in Table 13.2 and in Figure 13.12.

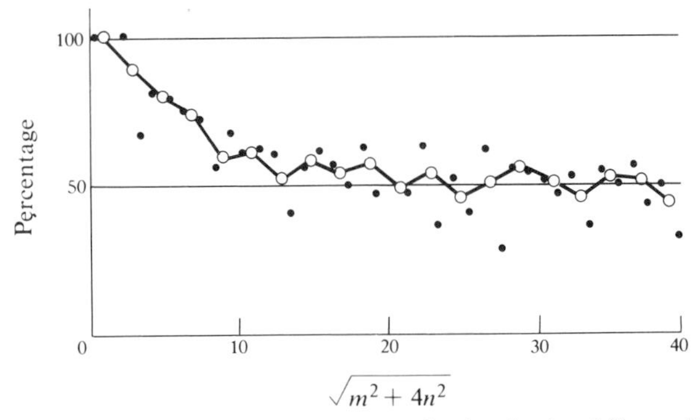

Figure 13.12 Relationship between the percentages of rational pairs of H_{mn} and B_{mn} and $\sqrt{(m^2 + 4n^2)}$.

Table 13.2 Percentage of rational pairs.

$m^2 + 4n^2$	*Total pairs*	*Rational pairs*	*Percentage*
0–2	3	3	100
2–4	18	16	89
4–6	30	24	80
6–8	42	31	74
8–10	54	32	59
10–12	70	43	61
12–14	82	43	52
14–16	90	52	58
16–18	110	59	54
18–20	95	54	57
20–22	90	44	49
22–24	80	43	54
24–26	78	37	47
26–28	84	43	51
28–30	72	40	56
30–32	82	41	50
32–34	72	34	47
34–36	76	40	53
36–38	51	26	51
38–40	16	7	44
40–42	1	0	0
Total	1296	712	55

As is clearly seen in Table 13.2 and Figure 13.12, the percentage of the rational pairs useful for determining d is close to 100 for smaller $\sqrt{(m^2 + 4n^2)}$ and tends to 50 as the order increases. This means that larger topographies are generally in isostatic equilibrium while smaller topographies are not necessarily so. Looking at the curve in Figure 13.12, $\sqrt{(m^2 + 4n^2)} = 20$ may be taken as the threshold value for 50% of the rational pairs over the total. From this, we see $\sqrt{2m^2} = 20$, or $m = 14.1$. Since a distance of 4000 km was taken to be 2π in the east–west direction, the wavelength of topography that corresponds to $m = 14.1$ is

$$4000 \div 14.1 = 280 \text{ km}$$

and the size of this topography is a half of the wavelength

$$280 \div 2 \approx 150 \text{ km}.$$

This value of 150 km corresponds to the regionality R of isostasy, for which Bowie found a value a little larger than 100 km.

Now in order to find the thickness of the Earth's crust, the rational pairs of H_{mn} and B_{mn} having lower orders $\sqrt{(m^2 + 4n^2)} < 20$ will be used.

Considering also the errors in the values of H_{mn} and B_{mn}, those pairs for which

$$H_{mn} > 50 \text{ m}$$

$$B_{mn} > 5 \text{ mgal}$$

will be used only. There are 15 such pairs and the error in d which is determined from these pairs is

$$\Delta d = \frac{1}{\sqrt{(m^2 + 4n^2)}} \sqrt{\left\{\left(\frac{\Delta B_{mn}}{B_{mn}}\right)^2 + \left(\frac{\Delta H_{mn}}{H_{mn}}\right)^2\right\}}, \qquad (13.10)$$

where $\Delta B_{mn} = 5$ mgal and $\Delta H_{mn} = 50$ m. Taking the weighted mean of the 15 values of d, a final result is obtained

$$d = 0.096 \pm 0.020$$

or in actual length

$$d = 4000 \times 0.096 \div 2\pi = 61 \pm 13 \text{ km}$$

because in this case, 4000 km was taken to be 2π.

The average height of topographies H_{00} in North America is 601 m and the average anomaly B_{00} is -70.9 mgal. By putting

$$-2\pi G\rho H_{00} = B_{00},$$

the mean density of the Earth's crust is found to be $\rho = 2.8$. The thickening of the Earth's crust due to H_{00} is

$$\frac{H_{00}\rho}{\rho' - \rho} = 5 \text{ km}.$$

Therefore the thickness of the Earth's crust corresponding to zero elevation is

$$61 - 5 = 56 \text{ km}.$$

Bowie's value for the depth of compensation based on Pratt's model was 120 km and the present value, 56 km, is just half of it, as should be the case.

As has been demonstrated, the isostatic thickness of the Earth's crust and the regionality could be found simply and directly by use of the Fourier series.

Of course, the calculated Bouguer anomaly $\Delta g_0''$ at a point does not exactly agree with that observed at that point. The difference of the two is called the isostatic anomaly, and it gives important information about deviations of the actual crustal structure from that of isostasy.

13.6 Gravity distribution along continental margins

Very characteristic distributions of gravity anomalies are generally observed along continental margins facing oceans. This fact was pointed out by F. R. Helmert of Germany in 1910. He noticed that a remarkably positive anomaly is observed on the continental side and a remarkably negative anomaly on the oceanic side of the margins.

Let us try to explain this characteristic distribution by assuming isostatic equilibrium along the continental margins. Continental margins may be idealized by a structure as shown in Figures 13.13 and 13.14. Compared with the normal oceanic crust, there is a positive mass plane (crustal density − seawater density) at a depth d and its compensating negative mass plane (crustal density − subcrustal density) at a depth kd. The value of gravity at a point due to a semi-infinite plane mass is proportional to the glancing angle of the edge of the plane seen from the

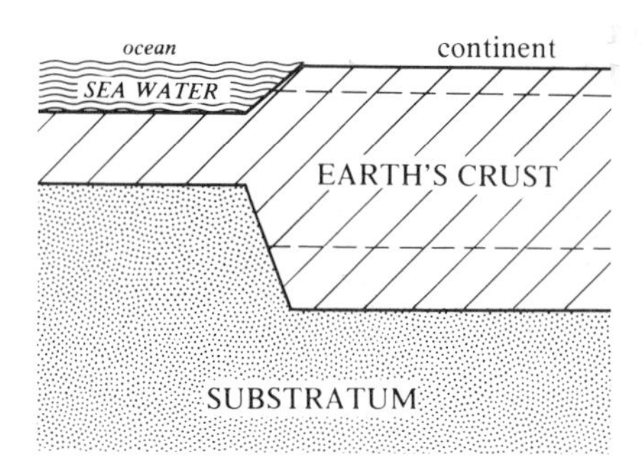

Figure 13.13 Isostasy along continental margins.

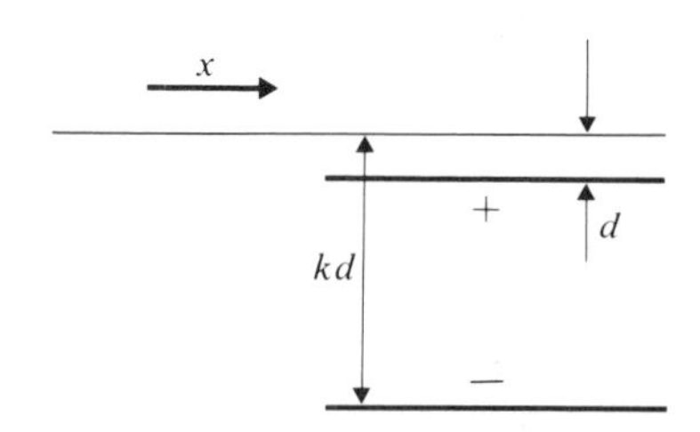

Figure 13.14 Schematic model of isostasy along continental margins.

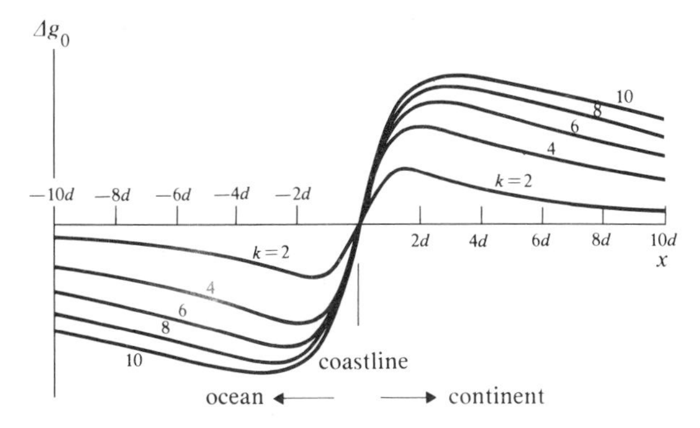

Figure 13.15 Gravity anomalies along continental margins at various distances from the coast line.

point, as was stated before. The calculated resultant gravity values due to the combined effects of the shallower negative mass plane and the deeper positive mass plane at various distances from the coast line are plotted in Figure 13.15 for various values of k. The characteristic distribution of the gravity anomalies along continental margins is not difficult to explain in this way assuming isostasy. Comparing the observed distribution of gravity anomaly with that calculated for various values of k, it will not be impossible to estimate the thickness of the Earth's crust. Actually, however, continental margins are usually associated with complicated geological structures, so that the estimation of the thickness of the Earth's crust in this way cannot be as accurate as desired.

13.7 The Earth's crust

As has been stated before, the same distribution of gravity is produced on the Earth's surface, whether Airy's isostasy with a thickness of the Earth's crust d is assumed or Pratt's isostasy with a depth of compensation $2d$. It is impossible, therefore, to decide which one of the models is closer to the truth from gravity data alone. For this decision, geophysical data other than gravimetric are needed. One of these is provided by studies of propagation of earthquake waves.

Suppose there is an Earth's crust of thickness d as shown in Figure 13.16. Let the velocity of propagation of earthquake waves in the crust be v and that in the subcrustal layer be V. V is larger than v. When an earthquake occurs at a point C on the surface of the crust, there are direct and indirect paths for earthquake waves to reach an observation station O at a distance of Δ from the origin. These two paths are shown in Figure 13.16: the direct one by a dotted line and the indirect one by

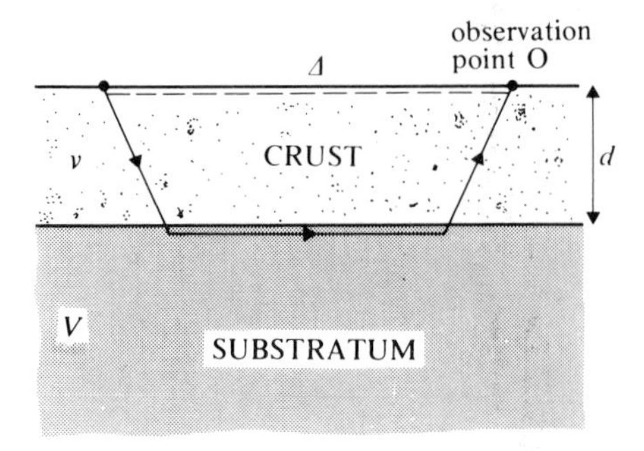

Figure 13.16 Direct and indirect paths of earthquake-wave propagation.

a full line. The direct wave is propagated through the crust with a velocity v. The indirect wave, on the other hand, hits the subcrustal layer and that portion of the waves, whose incidence angle corresponds to the critical angle of refraction, is propagated horizontally through the subcrustal layer with a velocity V and reaches the observation point O after being refracted into the crust again. If the distance Δ between the points C and O is small, the direct wave will arrive earlier, while if Δ is large, the indirect wave will arrive first. At a certain distance Δ_0, the two kinds of waves will arrive at the same time. According to the law of refraction, this distance Δ_0 is given by

$$\Delta_0 = 2d\sqrt{\left(\frac{V+v}{V-v}\right)}, \tag{13.11}$$

or

$$d = \frac{\Delta_0}{2}\sqrt{\left(\frac{V-v}{V+v}\right)}. \tag{13.12}$$

If, by observations, Δ_0, v and V can be found, the thickness of the Earth's crust d can be calculated. By using the probable values $\Delta_0 = 180$ km, $V = 8.0$ km s^{-1}, $v = 4.0$ km s^{-1}, d works out to be 53 km.

This phenomenon of dual paths of propagation of earthquake waves was discovered by A. Mohorovičić of Yugoslavia in 1909. The interface between the Earth's crust and the underlying subcrustal material at which the velocity of propagation of earthquake waves increases abruptly from v to V is called Mohorovičić discontinuity or simply Moho-discontinuity.

According to recent seismological observations, the depth of this discontinuity is not the same everywhere but is mostly 40–60 km. On the other hand, the thickness of Airy's crust is mostly 40–60 km from gravity data. It is natural therefore to consider that Mohorovičić's crust

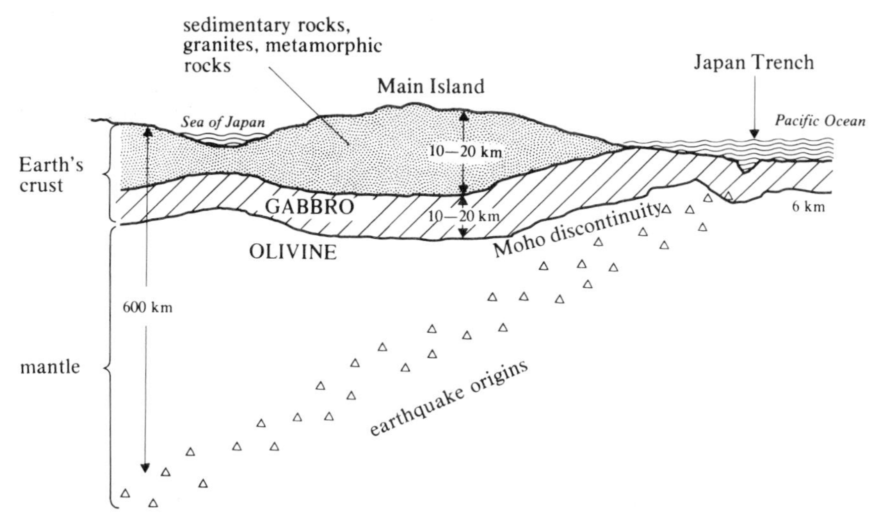

Figure 13.17 Rocks making up the Earth's crust and upper mantle across the E–W cross section of the northern part of Honshu (adapted from the 1961 work of H. Kuno).

and Airy's crust are the same thing. The subcrustal material existing below the Earth's crust is now called the **mantle** of the Earth. From petrological studies, the kinds of rocks of which the Earth's crust and the mantle are composed are believed to be as shown in Figure 13.17.

13.8 Polarity of isostasy

When the isostatic condition of the North American continent was studied in a previous section by means of the double Fourier series, two horizontal directions x and y were treated with equal weights. In other words, isostasy was considered to be an isotropic condition of the Earth's crust, having no directional character. But one can question whether this was a reasonable supposition, especially for studying the isostatic condition in an area where a mountain range is running in one particular direction.

A good example is provided in the Alpine area. The Alps run nearly in an east–west direction and the lines of equal Bouguer anomalies also follow the same direction. Let us see whether the relationships between the distributions of topography and Bouguer anomalies in this area differ according to direction and if they do, how. A rectangular area 600 km ($\lambda = 6.0°\text{E}–14.0°\text{E}$) in the east–west direction and 400 km ($\varphi = 46°\ 20'–50°\ 20'\text{N}$) in the north–south direction was taken and the distributions of topographical heights and Bouguer anomalies were expanded into a double Fourier series such as

$$h(x) = \sum_{m=0}^{12} \sum_{n=0}^{9} H_{mn} \begin{matrix}\cos\\ \sin\end{matrix} mx \begin{matrix}\cos\\ \sin\end{matrix} ny, \tag{13.13}$$

$$\Delta g_0''(x) = \sum_{m=0}^{12} \sum_{n=0}^{9} B_{mn} \begin{matrix}\cos\\ \sin\end{matrix} mx \begin{matrix}\cos\\ \sin\end{matrix} ny. \tag{13.14}$$

The order m was taken from 0 to 12 in the east–west direction and n from 0 to 9 in the north–south direction, considering that the area in question is not a square. In this analysis, there are 432 (i.e. $4 \times 12 \times 9$) pairs of H_{mn} and B_{mn} altogether. As was said before, the condition for isostasy requires that H_{mn} and the corresponding B_{mn} be different in algebraic sign. For each pair of m and n, there are four different components $\sin mx \sin ny$, $\sin mx \cos ny$, $\cos mx \sin ny$, $\cos mx \cos ny$.

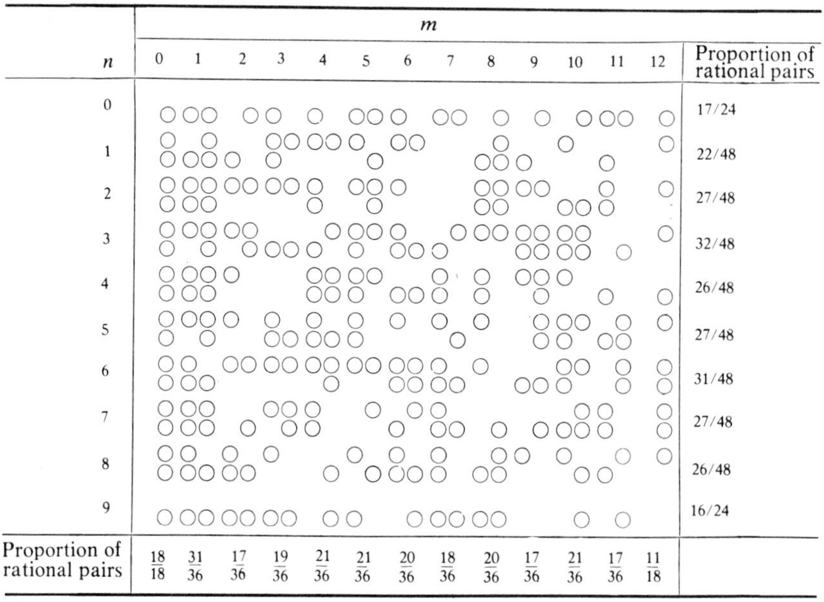

n \ m	0	1	2	3	4	5	6	7	8	9	10	11	12	Proportion of rational pairs
0														17/24
1														22/48
2														27/48
3														32/48
4														26/48
5														27/48
6														31/48
7														27/48
8														26/48
9														16/24
Proportion of rational pairs	18/18	31/36	17/36	19/36	21/36	21/36	20/36	18/36	20/36	17/36	21/36	17/36	11/18	

Figure 13.18 Rational pairs indicated by circles.

sin mx sin ny	cos mx sin ny
sin mx cos ny	cos mx cos ny

Figure 13.19 Four smaller squares corresponding to four different components.

In Figure 13.18, a square belonging to a pair of m and n is divided further into four smaller ones, each corresponding to one of the four different components as indicated in Figure 13.19. If H_{mn} and B_{mn} rationally differ in algebraic sign, a circle is drawn in Figure 13.18, in the smaller square which corresponds to its order and component. As is seen in the figure, the abundance of the circles is remarkably different in the east–west and north–south directions. For $n = 0, 1, m = 0$–12, the percentage of the rational pairs with circles over the total is

$$\frac{17 + 22}{24 + 48} = \frac{39}{72} = 54\%,$$

while for $m = 0, 1, n = 0$–9, the percentage is

$$\frac{18 + 31}{18 + 36} = \frac{49}{54} = 91\%.$$

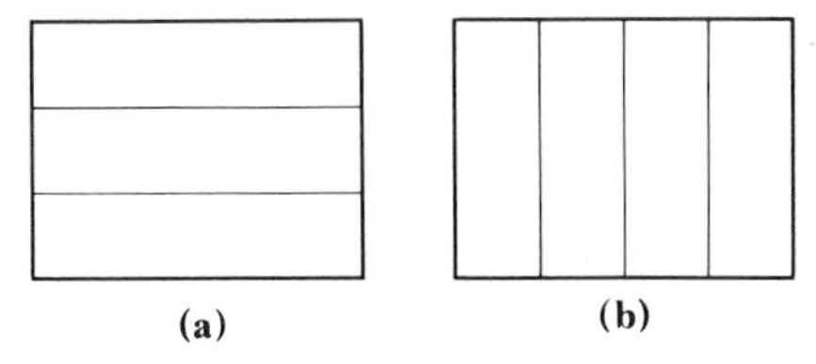

Figure 13.20 Topographical patterns varying in (a) the north–south and (b) the east–west directions.

This shows that large topographies which vary in the north–south direction, as in Figure 13.20a, are mostly in isostatic equilibrium, while those varying in the east–west direction, as in Figure 13.20b, are not. This phenomenon invites an interesting further study as to how the isostatic condition is brought about in relation to mountain-building processes. In the Alps, the variations of topographies in the north–south direction must be closely related to its primary mountain-building process. Those in the east–west direction were produced in a more or less accidental way by erosions or minor geological movements which have little to do with isostasy.

13.9 Process of development of isostasy

In connection with the problem of how isostasy is achieved under various circumstances, the upheaval of the ground in the Scandinavian area is often cited. The beach lines at the time of the latest glacial period can be traced now at elevations of a few hundred metres above the

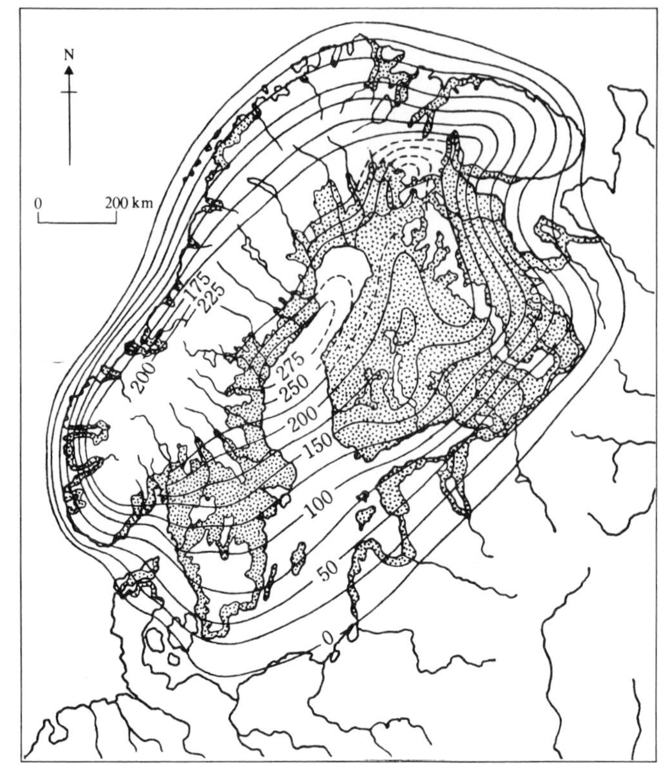

Figure 13.21 Lines of equal uplift (m) in Scandinavia.

present sea level, the highest one being at 275 m. The lines of equal upheaval shown in Figure 13.21 indicate that this area has been uplifted in a form of a shield since the disappearance of the glacial ice which had covered this area before. If the uplift is considered to be due to a recovery of an original elastic depression of the ground which had been produced by the weight of the ice, the depression must have been at least 275 m, and the ice necessary for producing such a large depression must have been 6000–7000 m in thickness. This value is unbelievably large.

In contrast to this, the uplift can more easily be explained as follows with the help of the idea of isostasy. Referring to Figure 13.22, let us assume that the Earth's crust is in isostatic equilibrium whether there is ice (Figure 13.22a), or not (Fig. 13.22b) on the surface. In Figure 13.22a, the isostatic condition is

$$\rho_1 h_1 + \rho_2 h_2 = \rho_3 D_1$$

and in Figure 13.22b, it is

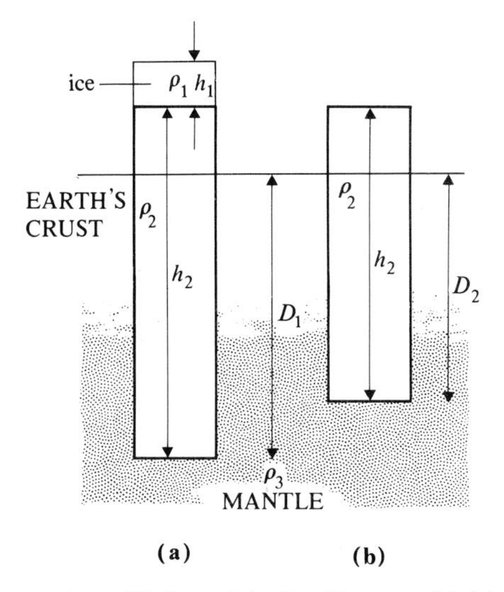

Figure 13.22 Isostatic equilibrium of the Earth's crust with (a) an ice sheet and (b) without an ice sheet.

$$\rho_2 h_2 = \rho_3 D_2.$$

Subtracting the second equation from the first, we obtain

$$\rho_1 h_1 = \rho_3(D_1 - D_2) \tag{13.15}$$

or

$$h_1 = \frac{\rho_3}{\rho_1}(D_1 - D_2).$$

$(D_1 - D_2)$ is simply the amount of uplift of the ground. By putting $D_1 - D_2 = 275$ m, $\rho_1 = 0.9$, $\rho_3 = 3.0$, the thickness of the ice sheet h_1 comes out to be

$$h_1 = \frac{3.0}{0.9} \times 275 \approx 920 \text{ m}$$

and this value is easy to accept. This shows that isostatic recovery was the probable cause of the uplift in Scandinavia.

Let us proceed now to see what will happen to the Earth's crust if its surface mass is decreased by erosion or is increased by additional

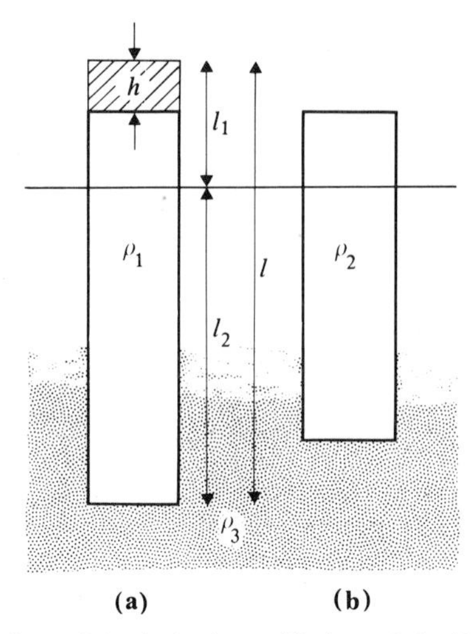

Figure 13.23 Comparison of the isostatic equilibrium of the Earth's crust (a) before erosion and (b) after erosion.

sedimentation. Figure 13.23 shows a comparison of the isostatic equilibrium of the Earth's crust before (13.23a) and after (13.23b) erosion. In Figure 13.23a, the isostatic condition is

$$\rho_2(l_1 + l_2) = \rho_3 l_2,$$

or

$$\frac{l_1}{l_2} = \frac{\rho_3 - \rho_2}{\rho_2}.$$

From this, it is seen that

$$\frac{l_1}{l_1 + l_2} = \frac{\rho_3 - \rho_2}{\rho_3}, \tag{13.16}$$

which means that the thickness of the crust l_1, coming out from the mantle is $(\rho_3 - \rho_2)/\rho_3$ times the total thickness. In Figure 13.23b, after a thickness of h is taken away by erosion, the total thickness of the crust is $(l_1 + l_2 - h)$. Then the part of the crust coming out from the mantle is

$$(l_1 + l_2 - h)\,\frac{\rho_3 - \rho_2}{\rho_3}.$$

Before erosion, this was

$$(l_1 + l_2)\frac{\rho_3 - \rho_2}{\rho_3},$$

so that the lowering of the Earth's surface is $h(\rho_3 - \rho_2)/\rho_3$. If a thickness of 1 km is considered to have been eroded away, the final lowering of the Earth's surface is smaller than this and is

$$1 \times \frac{3.0 - 2.7}{3.0} = 0.1 \text{ km}.$$

The remaining $1.0 - 0.1 = 0.9$ km, which was taken away by erosion is cancelled by a rising of the crust according to isostasy. In other words, if the surface was finally lowered by 100 m by erosion, the part which had been at a depth of 900 m will come up to be seen on the new crustal surface. Plutonic rocks which are believed to have been formed at great depths are now seen exposed at many places on the Earth's surface. If they were formed at a depth of H, the total amount of erosion of the Earth's surface which will be needed to bring them to the surface by the mechanism of isostasy is $9H$.

The reverse process of erosion is sedimentation. In Figure 13.24, the sea water, a sedimentary deposit, the Earth's crust and the mantle are regarded to be in isostatic equilibrium. Before sedimentation,

$$\rho_1 h_1 + \rho_3 h_3 + \rho_4 h_4 = \rho_3 D,$$

and after sedimentation

$$\rho_1 h_1' + \rho_2 h_2' + \rho_3 h_3 + \rho_4 h_4' = \rho_3 D.$$

Subtracting the first equation from the second,

$$\rho_1(h_1' - h_1) + \rho_2 h_2' + \rho_4(h_4' - h_4) = 0,$$

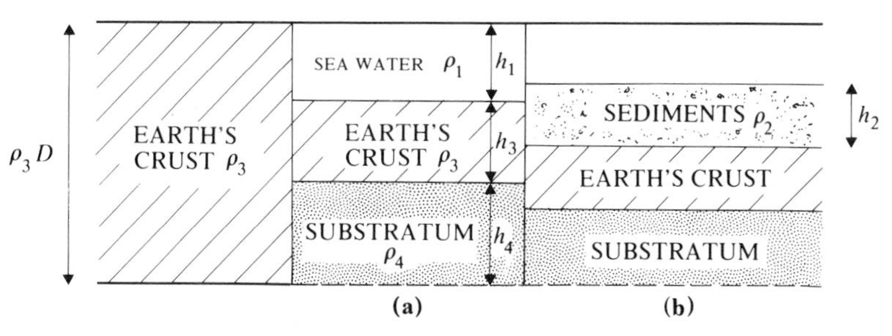

Figure 13.24 Isostatic equilibrium of the Earth's crust (a) before sedimentation and (b) after sedimentation.

and from geometry

$$(h_1' - h_1) + h_2' + (h_4' - h_4) = 0.$$

From these two equations, it is seen that

$$h_1' - h_1 = -\frac{\rho_4 - \rho_2}{\rho_4 - \rho_1} h_2'. \tag{13.17}$$

Since $(\rho_4 - \rho_1)$ and $(\rho_4 - \rho_1)$ are both positive, $(h_1' - h_1)$ is negative. This means that the sea bottom becomes shallower by sedimentation. By putting $\rho_1 = 1.0$, $\rho_2 = 1.5$ and $\rho_4 = 3.0$, we see

$$\frac{\rho_4 - \rho_2}{\rho_4 - \rho_1} = \frac{1.5}{2.0} = 0.75,$$

so that

$$h_1' - h_1 = -0.75 h_2.$$

If the sediments are of h_2' in thickness, the sea will not become shallower by that amount but by only 0.75 times of the thickness. For instance, if there is sedimentation of 1000 m in thickness, the sea will become shallower by 750 m. The difference 1000 − 750 = 250 m must have been produced by a sinking of the sea bottom according to isostasy. It is important to notice that the sinking of the sea bottom by isostasy is smaller than the thickness of sediments.

In various parts of the world, very large, trough-like sedimentary deposits amounting to several kilometres in thickness are seen. These features are called **geosynclines**. The sea bottom must have sunk enormously to produce such thick structures but, as shown above, the weight of sediments alone cannot do this. In order to produce a geosyncline, some other agent must be considered to have caused the

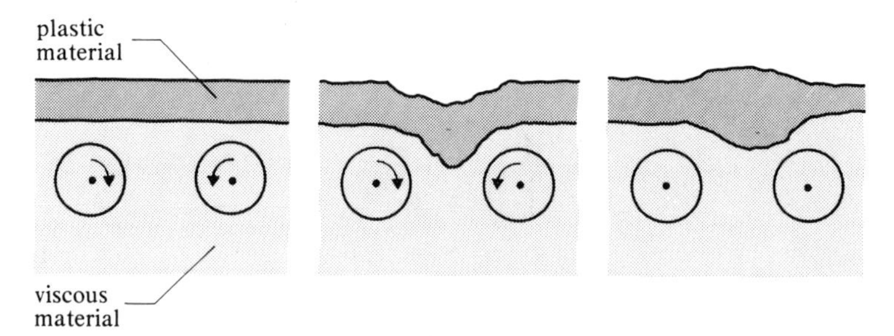

Figure 13.25 Griggs' experiment.

depressions. Griggs (1939) made a useful experiment in this connection. He put a plastic layer on a very viscous liquid as shown in Figure 13.25. Two cylinders which can be rotated in the directions of arrows are put in the viscous liquid. The viscous liquid is dragged down by the rotation of the cylinders and the surface of the plastic layer is depressed and so becomes thicker. When the rotation of the cylinders is stopped, the plastic layer begins to rise by buoyancy. This was considered to be a small-scale model of an isostatic recovery. Formation of a geosyncline can be understood by a mechanism similar to that seen in Griggs' experiment.

13.10 Mountains and isostasy

There is no doubt that major crustal topography, such as mountains and oceans, are isostatically balanced. There are two alternative ways for explaining how equilibrium is achieved.

One is to consider that mountains are formed first as added loads near the upper surface of the Earth's crust by some mechanism and that this part of the crust was depressed by its weight into the mantle until isostatic equilibrium is attained. The other interpretation is the reverse of this. A downward bulging of the Earth's crust into the mantle is formed first by some mechanism and that part of the crust rises by buoyancy until the isostatic balance is attained. In short, the problem is whether mountains are the causes or the results of isostasy. Griggs' experiment is useful in this connection. This problem is elusive in nature, but it can be discussed to some extent mathematically in the following way.

Suppose there is an isolated unit mountain on the Earth's surface which is otherwise uniform in thickness. Let $\varphi(x)$ be the undulation form of the Moho-discontinuity produced by the weight of the unit mountain. Now, if instead of an isolated unit mountain there is an undulation $H(x)$ on the upper surface of the Earth's crust, the resulting undulation $D(x)$ at the Moho-discontinuity is given by

$$D(x) = \int_{-\infty}^{\infty} H(x+\alpha)\varphi(\alpha)\, \mathrm{d}\alpha. \tag{13.18}$$

On the other hand, if there is a unit bulging of the Earth's crust into the mantle, the resulting undulation of the surface topography is considered to be $\psi(x)$. If, instead of an isolated down-bulging of the Earth's crust, there is an undulation $D(x)$ at the Moho-discontinuity, the resulting undulation of the surface topography is given by

$$H(x) = \int_{-\infty}^{\infty} D(x + \alpha)\psi(\alpha)\, d\alpha. \tag{13.19}$$

The two equations given above are integral equations for finding $\varphi(x)$ and $\psi(x)$ from $H(x)$ and $D(x)$.

$$F(x) = \Sigma(A_m \cos mx + B_m \sin mx),$$

$$f(x) = \Sigma(a_m \cos mx + b_m \sin mx),$$

$$\phi(x) = \Sigma(\alpha_m \cos mx + \beta_m \sin mx),$$

and if

$$F(x) = \int_{-\infty}^{\infty} f(x + \alpha)\phi(\alpha)\, d\alpha,$$

then the following relations can be proved to exist

$$\alpha_m = \frac{1}{\varepsilon\pi} \frac{a_m A_m + b_m B_m}{a_m{}^2 + b_m{}^2}$$

$\varepsilon = 2$ for $m = 0$, and $\varepsilon = 1$ for $m \geq 1$, and

$$\beta_m = \frac{1}{\pi} \frac{b_m A_m - a_m B_m}{a_m{}^2 + b_m{}^2}, \tag{13.20}$$

among the Fourier coefficients. If $H(x)$ and $D(x)$ are given, the weight functions $\varphi(x)$ and $\psi(x)$ can be calculated from them by using the above expressions. The two weight functions correspond to the cases when the mountain is considered to be the cause and when it is considered to be the result, respectively. $H(x)$ is the surface topography and $D(x)$ is the undulation of the Moho-discontinuity. Both can be found from observations or from gravity interpretations.

This method was applied to the average profiles of surface topography and the undulation of the Moho-discontinuity in an east–west direction across the United States found from gravity data. The forms of the weight functions $\varphi(x)$ and $\psi(x)$ calculated are compared in Figure 13.26. The curve for $\psi(x)$ is seen to have a much more reasonable form than $\varphi(x)$ and the conclusion can be drawn that the downwarping of the

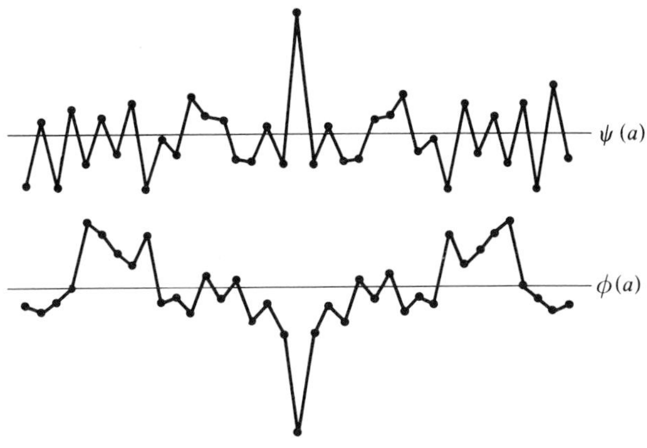

Figure 13.26 Comparison of weight functions of $\psi(x)$ and $\phi(x)$.

Moho-discontinuity is likely to be the primary origin of isostasy. All this shows that by some process in the mantle, a downwarping of the crust was produced, first forming a thick geosynclinal structure and this feature rose at a later stage by the action of isostasy. It is a noteworthy fact that high mountain ranges in the world are generally synclinal in geological structure and they are sedimentary rocks of great thicknesses having concave curvatures upwards.

13.11 Thickness of the Earth's crust and regionality

By using the Fourier series, the thickness of Airy's crust and regionality of isostasy can easily be derived as described in a previous section. Some of the results found for various parts of the world are given in Table 13.3.

Table 13.3 Thickness of Airy's crust d and regionality R in various parts of the world.

Place	d	R	R/d
Middle Japan	50 km	85 km	1.7
USA	61 km	145 km	2.4
West East Indies	36 km	60 km	1.7
East East Indies	29 km	80 km	2.8
Korea	30 km	85 km	2.7

The values of d and R differ considerably for various areas, but the ratio R/d does not vary so widely and it is around 2–3, irrespective of the values of d and R. This is an interesting and important fact in relation to the mechanical strength of the Earth's crust.

13.12 The magnitude of topographic variations and the depth of isostasy

In the above sections, it has been assumed that there is only one Earth's crust and various topographies are supported by the buoyancy forces acting along its lower surface. It is more reasonable to assume that the masses of relatively smaller topographies are compensated at shallower depths than those of larger topographies.

It has already been shown that the thickness of the Earth's crust d is given by

$$d = -\frac{1}{\sqrt{(m^2+n^2)}}\log_e\left(-\frac{B_{mn}}{2\pi G\rho H_{mn}}\right). \qquad (13.21)$$

Using gravity data from North America, the overall average of d which was calculated by taking all m and n was found to be 56 km. If the value of d is calculated separately according to the wavelength λ of the topography, d is found to be dependent on λ as shown in Figure 13.27. The points in the figure are approximately arranged on two straight lines. For topography for which $\lambda = 250$–300 km, the depth is uniformly around 60 km. For shorter wavelength topography, d varies according to a relation

$$d = \frac{1}{3}\frac{\lambda}{2}.$$

It is likely that shorter wavelength topography are not supported by the crust–mantle buoyancy, but are supported by the density contrasts of various crustal layers.

Recent artificial seismic wave exploration has revealed that the

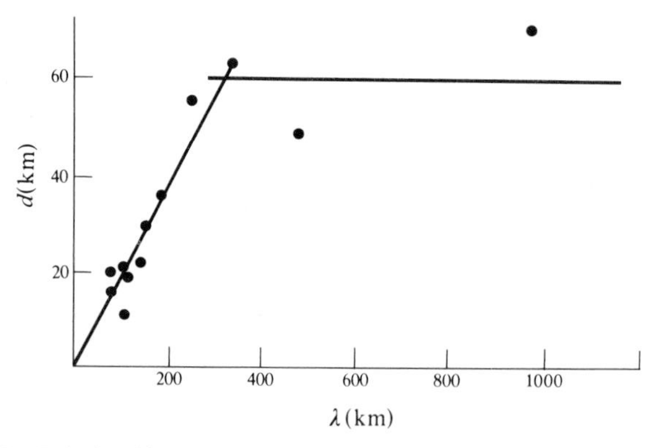

Figure 13.27 Relationship between the values of d and the wavelength λ topography.

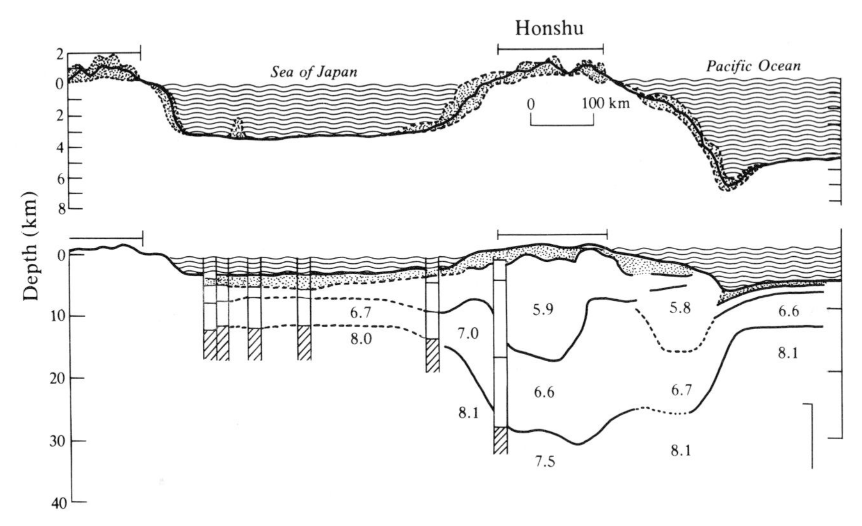

Figure 13.28 East–west cross-section of the northern part of Honshu, Japan. The numbers in the lower figure refer to the velocities of seismic waves (P).

Earth's crust is not a single plate of uniform density, but is composed of several layers with various velocities of earthquake waves. For instance, T. Yoshii showed in 1977 that the east–west profile of the northern part of Honshu is as shown in Figure 13.28. From the velocity distribution of seismic waves, Yoshii divided the Earth's crust into three layers, A, B and C. The layer A in which the velocity is the smallest has shorter wavelength undulations at its bottom and the deeper layer, B, has longer wavelength undulations. This may perhaps explain the dependence of d on the wavelength of topography to be balanced.

13.13 Earthquake magnitude and isostasy

The idea of isostasy has had a large influence on our understanding of the energy of earthquakes, especially in relation to their magnitudes. The concept of earthquake magnitude M was proposed by C. F. Richter in 1935. Among various things which have been found about the magnitude, two are of special interest:

(a) there is likely to be an upper limit of M at $M = 8.6$;
(b) the energy of an earthquake of magnitude M is given by

$$\log E = 11.8 + 1.5M.$$

From these, the energy of an earthquake of the largest magnitude 8.6 is

$$\log E = 11.8 + 1.5 \times 8.6 = 24.7,$$

$$\begin{aligned} E &= 5 \times 10^{24} \times 10^{-7}\ \text{J} \\ &= 5 \times 10^{17}\ \text{J}. \end{aligned}$$

On the other hand, according to studies of isostasy, the thickness of the Earth's crust d is around 60 km and the regionality R is about three times d. The maximum mechanical energy that can be stored up in a volume of the Earth's crust of this dimension may be identified to be the maximum energy of an earthquake. The energy that can be stored up in a unit volume of the Earth's crust is

$$\begin{aligned} e &= \tfrac{1}{2} \times (\text{elastic constant}) \times (\text{ultimate strain})^2 \\ &= \tfrac{1}{2} \times 5 \times 10^{11} \times (10^{-4})^2 \\ &= 3000\ \text{erg} \\ &= 3 \times 10^{-4}\ \text{J}. \end{aligned}$$

If this energy is stored up in a volume of the Earth's crust of $60 \times 180 \times 180\ \text{km}^3$, the total energy will amount to

$$\begin{aligned} E &= 3000 \times 60 \times 180 \times 180 \times 10^{15} \\ &= 6 \times 10^{24}\ \text{erg} \\ &= 6 \times 10^{17}\ \text{J}, \end{aligned}$$

which agrees with the energy of an earthquake of the largest magnitude 8.6. Here a close relation is seen between the earthquake and isostasy.

Utsu (1955) pointed out that the after-shock area A of a large earthquake of magnitude M is given by

$$\log A = M + 6.$$

Putting $M = 8.6$ in this equation

$$\log A = 8.6 + 6 = 14.6.$$

If this area is considered to be a square in form, with side a, then

$$\begin{aligned} \log a^2 &= 14.6 \\ \log a &= 7.3 \\ a &= 2 \times 10^7 = 200\ \text{km}, \end{aligned}$$

which is very close to the average value of regionality R. Moreover, it is a fact that seismic activities in various parts of Japan are correlated to a distance of 200 km but no more beyond that (Tsuboi 1950b). This value of 200 km agrees with the regionality R also. All these things can be understood as different indications of the finite mechanical strength of the Earth's crust.

References and further reading

Airy, G. B. 1855. On the computation of the effect of the attraction of mountain masses as disturbing the apparent astronomical latitude of stations in geodetic surveys. *Phil Trans R. Soc., Lond.* **145**, 101.

Born, A. 1932. *Isostasie und Scheremessungen*. Berlin: Springer.

Bowie, W. 1917. *Investigation of gravity and isostasy*. Spec. Pub. USCGS 40, 1.

Bowie, W. 1927. *Isostasy*. New York: Dutton.

Bowie, W. 1942. *Isostatic investigations and data for gravity stations in the United States established since 1915*. Spec. Pub. USCGS 99, 1.

Dutton, C. E. 1889. *On some of the great problems of physical geology*. Bull. Phil Soc., Washington, no. 11.

Griggs, P. 1939. A theory of mountain building. *Am. J. Sci.* **237**, 611.

Hayford, J. H. 1909. *The figure of the Earth and isostasy from measurements in the United States*. Spec. Pub. USCGS, 1.

Hayford, J. H. and W. Bowie 1912. *The effect of topography and isostatic compensation upon the intensity of gravity*. Spec. Pub. USCGS 10, 1.

Heiskanen, W. 1924. *Untersuchungen uber Schwerkraft und Isostasie*. Veroff. Finn. Geod. Inst., no. 4.

Heiskanen, W. 1936. Das Problem der Isostasie. *Handb. Geophys.* **1**, 4.

Mohovičić, A. 1910. *Das Beben vom 8, X, 1909*. Jahrb. Met. Obs. fur das Jahr 1909, Zagreb.

Pratt, J. H. 1855. On the attraction of the Himalayan mountains and of the elevated regions beyond them upon the plumb-line in India. *Phil Trans R. Soc., Lond.* **145**, 53.

Pratt, J. H. 1859. ibid. **149**, 745, 749.

Pratt, J. H. 1871. ibid. **161**, 335.

Tsuboi, C. 1938. Hypotheses of Airy and Pratt (in Japanese). *J. Seis. Soc., Japan* **10**, 109.

Tsuboi, C. 1940a. The direct and indirect methods for determining the thickness of the isostatic Earth's crust. *Geophys. Notes*, no. 4.

Tsuboi, C. 1940b. Isostasy and maximum earthquake energy. *Proc. Imp. Acad., Japan* **16**, 449.

Tsuboi, C. 1942. Relation between the gravity anomalies and the corresponding subterranean mass distribution (VII). *Bull. Earthquake Res. Inst.* **20**, 30.

Tsuboi, C. 1950a. Thickness of the isostatic Earth's crust in various parts of the United States of America. *Geophys. Notes*, no. 5.

Tsuboi, C. 1950b. Dependence of the isostatic depth on the horizontal scale of the topographies to be compensated. *Geophys. Notes*, no. 6.

Tsuboi, C. 1956. Earthquake energy, earthquake volume, aftershock area and strength of the Earth's crust. *J. Phys. Earth* **4**, 63.

Tsuboi, C., T. Kaneko, S. Miyamura and T. Yabasi 1939. Relation between the gravity anomalies and the corresponding mass distribution. Isostasy in the United States of America. *Bull. Earthquake Res. Inst.* **17**, 385.

Utsu, T. 1955. A relationship between the aftershock area and the energy of the main shock (in Japanese). *J. Seis. Soc., Japan* **7**, 233.

14 Gravity at sea

14.1 Oceans and continents

Accurate measurement of gravity values at sea was made possible by Vening Meinesz in the 1920s. His instrument for use on submarines was described in Chapter 2, together with other instruments which have since been developed for the purpose. At present, gravity values can be measured continuously on board sailing surface ships with an accuracy of 1 mgal. From these measurements, a number of interesting geophysical facts have been found about the structure of the sea bottom Here the word 'sea' is used in a specific sense. It does not simply mean those parts on the Earth's surface which are covered by sea water. In this specific sense, shallow seas which border lands are not real seas, not to speak of small bays along coastlines. When the distribution of Bouguer gravity anomalies around Tokyo Bay (occupying an area of 50 km × 30 km) is plotted on a map, the contour lines on land on both sides of the bay are seen to continue smoothly across it as shown in

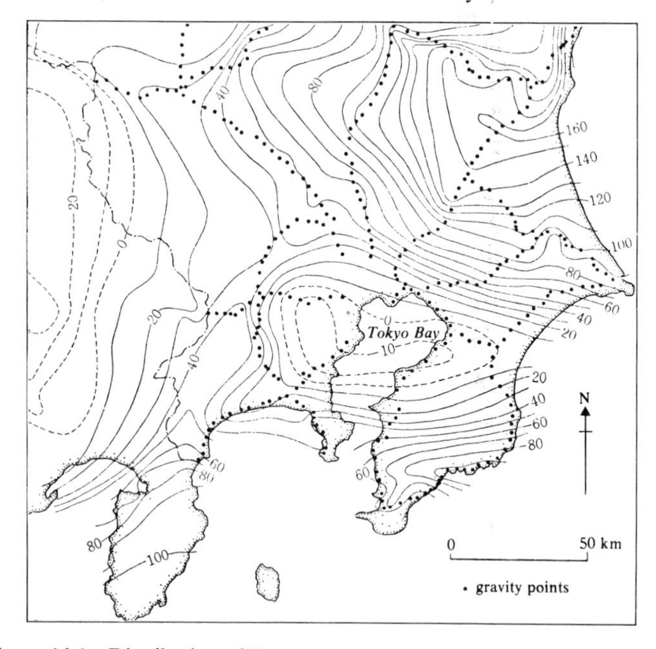

Figure 14.1 Distribution of Bouguer anomalies around Tokyo Bay, Japan.

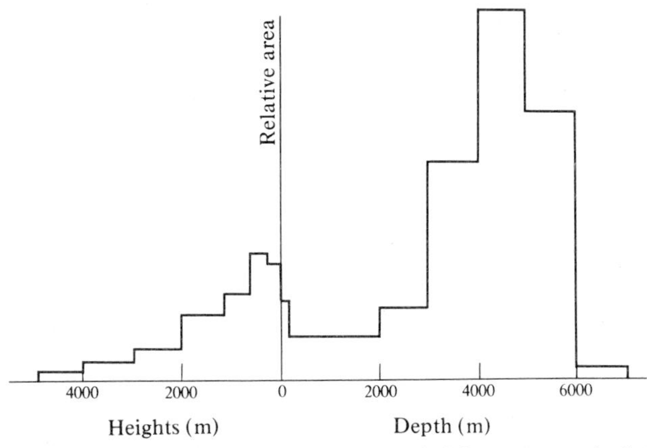

Figure 14.2 Relative areas which have various land heights and oceanic depths on the Earth's surface.

Figure 14.1. This indicates that the bay does not differ much in geological structure from its neighbouring land areas. It is not a sea in this sense, but is merely a low concave part of the Earth's surface which is covered by sea water.

Figure 14.2 shows the percentage of areas on the surface of the Earth according to heights and depths. Two distinct maxima are seen at $h = 0$ and $h = -5000$ m. If the topography on the Earth's surface had been produced by random processes, such a distribution would not have resulted. The two maxima in Figure 14.2 represent two fundamentally different units of the Earth's surface structure, continents and oceans. The bottoms of the Pacific and Atlantic oceans are the representatives of the oceanic floor. The oceans are of course not uniformly 5000 m in depth everywhere and there are great mountain ridges, wide flat basins and high sea-mounts on their floor. Gravity measurements are useful for finding the difference between the oceanic and continental structure of the Earth's crust and also for investigating the mechanism by which various characteristic features of ocean bottoms were produced.

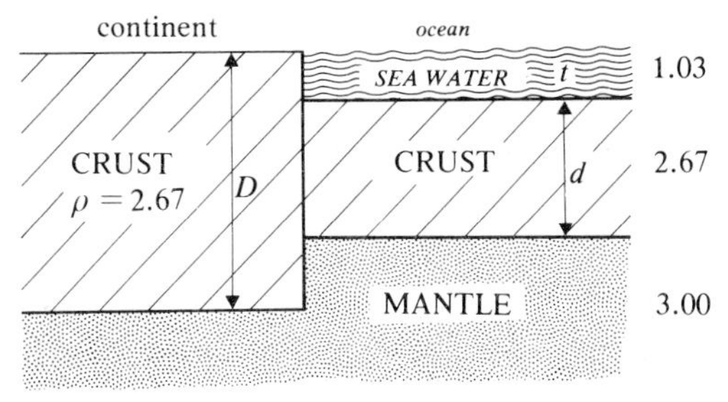

Figure 14.3 A simplified model of the isostatic equilibrium of a continent and ocean.

In a very simplified model, continents and oceans can be considered to be as shown in Figure 14.3 and to be in isostatic equilibrium. With reference to the figure, the condition for the isostatic equilibrium

$$2.67D = 1.03t + 2.67d + 3.00(D - t - d), \tag{14.1}$$

where 1.03, 2.67 and 3.00 are the densities of the sea water, the Earth's crust and the mantle respectively. From this, we obtain

$$d \approx D - 6t.$$

If $D = 40$ km, $t = 5$ km, it follows that $d = 10$ km. According to this model, the Earth's crust under the oceans is much thinner than the continental crust. Δg_0 over the oceans is nearly zero for the most part and it is by this observation that the isostatic model is justified.

There are many places on the ocean bottom where its structure differs from one in isostatic equilibrium and large gravity anomalies are observed. In the case of Figure 14.4a, there is a rise of the ocean bottom. If this rise is not very large in extent and is supported by the elasticity of the crust, not by isostasy, then Δg_0 above the rise will be positive. If the Δg_0 values are corrected for the attraction of the mass of the rise, Bouguer anomalies $\Delta g_0''$ will be uniformly positive across the rise, showing no spatial variation. Figure 14.4b shows the case in which the ocean bottom is flat but the Earth's crust under it is thick owing to its downward bulge into the underlying mantle. In this case, both Δg_0 and $\Delta g_0''$ will show a similar distribution of low values. In the case of Figure 14.4c, on the other hand, where the ocean floor is flat and the Earth's crust is thin, both Δg_0 and $\Delta g_0''$ will show similar distributions of relatively high values. There are thus two different types in the comparative distributions of Δg_0 and $\Delta g_0''$ on oceans. One is due to the topography of the ocean bottom and the other is due to the undulation of the interface

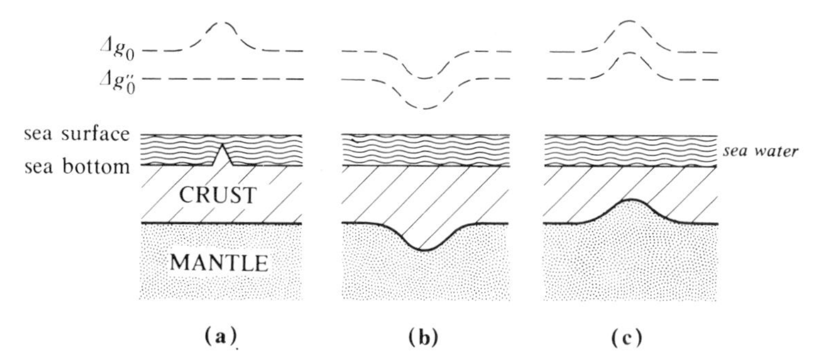

Figure 14.4 Three types of distribution of Δg_0 and $\Delta g_0''$: (a) when there is a rise in the ocean bottom; (b) when the ocean bottom is flat, but the Earth's crust under it is thick; and (c) when the ocean floor is flat and the Earth's crust is thin.

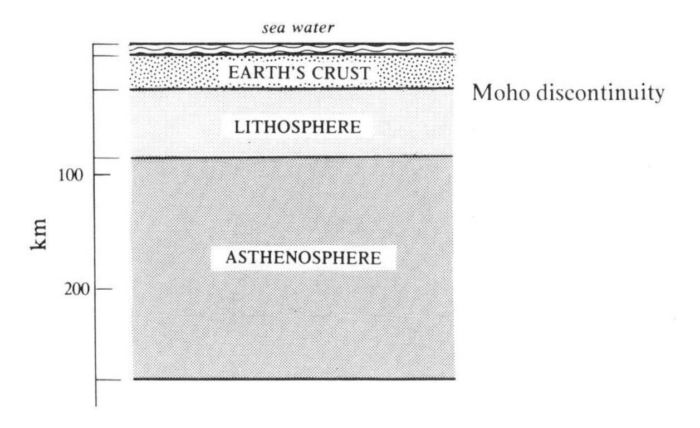

Figure 14.5 A more complicated model of the upper part of the Earth.

between the oceanic crust and the underlying mantle. In actual cases, these effects occur in combination. Ocean-bottom topography is observed by soundings and their attractions can be calculated. If the observed Δg_0 is corrected for the attractions of topographies and if it cannot be accounted for by this, the residual must be explained by an undulation of the interface between the oceanic crust and the mantle.

Various studies other than gravimetric, especially seismic, are being made to derive the structure of the ocean bottom, and many cases have been found in which the simple model consisting of the sea water, the crust and the mantle does not fit well with observed facts and a more complicated structure, as shown in Figure 14.5, has been proposed. This model consists of the sea water, the crust, the lithosphere, the asthenosphere and the mantle. This model assumes the existence of a weak layer, called the asthenosphere. The problem of how the equilibrium can be held in this structure remains to be studied.

14.2 Gravity above ocean ridges

An ocean ridge is a very long chain of mountains extending over several thousand kilometres on the ocean floor. The Mid-Atlantic Ridge, with a length of more than 10 000 km and an east–west width of 2000 km, rising more than 3000 m above the floor of the Atlantic ocean, is a magnificent example of this.

A group of scientists from Lamont–Doherty Geological Observatory, Columbia University, made various kinds of geophysical surveys across the Mid-Atlantic Ridge in an area around 32°N and 40°W. They combined the results of gravimetric, seismic and other observations and came to the following conclusions.

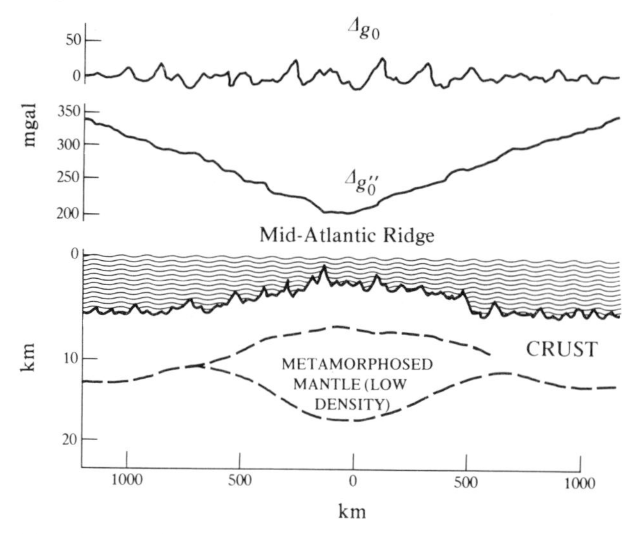

Figure 14.6 Δg_0, $\Delta g_0''$ and the proposed structure across the Mid-Atlantic Ridge.

(a) The top of the ridge rises about 3000 m from the general floor of the Atlantic Ocean.
(b) There are topographical irregularities on the slope of the ridge amounting to several hundred metres in amplitude.
(c) Δg_0 across the ridge is nearly zero on average but their distribution is superposed by irregularities of the order of 10 mgal. These irregularities can be accounted for by the attractions of the irregular topography.
(d) When topographical corrections are applied to Δg_0, the resulting $\Delta g_0''$ shows little irregularities. $\Delta g_0''$ above the ridge is 200 mgal smaller than on both sides of the ridge.
(e) Seismic surveys across the ridge do not show any indication that the Earth's crust bulges downward into the mantle.
(f) Combining all the above findings, it is concluded that the mantle under the ridge is metamorphosed and its density is abnormally low.

The structure of the profile across the ridge as proposed by the Columbia team is shown in Figure 14.6.

14.3 Gravity over ocean trenches

Vening Meinesz was the first to observe the gravity distribution over oceanic trenches. In 1929, he made an extensive gravimetric voyage on board a Dutch submarine from the Netherlands to the Mediterranean Sea, Indian Ocean, East Indies, Pacific Ocean and Atlantic Ocean. The

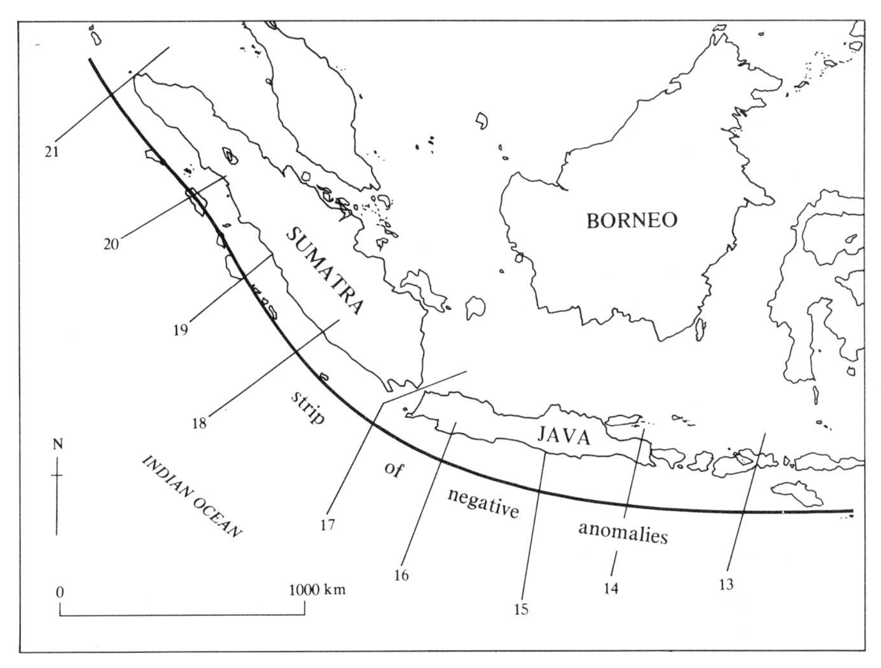

Figure 14.7 Strip of negative gravity anomalies along the East Indies.

most remarkable thing in his results is the discovery of an extensive belt or strip of negative gravity anomalies over the deep trench running along the ocean side of Sumatra, Java, Sulawesi (Celebes) and Cermus. The belt of negative anomalies is shown in Figure 14.7. At some points, the anomaly exceeds −150 mgal. In order to explain this remarkable feature, Vening Meinesz assumed a large downward bulge of the crust into the mantle. This was a revolutionary discovery in connection with the mechanism of trench formation.

In 1939, the present author applied his Fourier series method for interpreting the gravity-anomaly distribution found by Vening Meinesz in the East Indies. Figure 14.8 shows one of the results of calculations which was applied to Meinesz Profile 17 across the island arc and two shapes of the bottom surface of the crust, h and h', are compared. One of them, h' was calculated on the assumption of isostasy and the other one, h, calculated from the distribution of gravity anomalies by the Fourier series method. The thickness of the crust varies generally in accordance with isostasy but a remarkable downward bulge ($h - h'$) is clearly seen amounting to almost 10 km. It should be noted that this downward bulge is not located immediately beneath the trench but between the island arc and the trench.

Similar belts of remarkable negative gravity anomalies have been found over the West Indies Trench (by H. H. Hess in 1933) and over

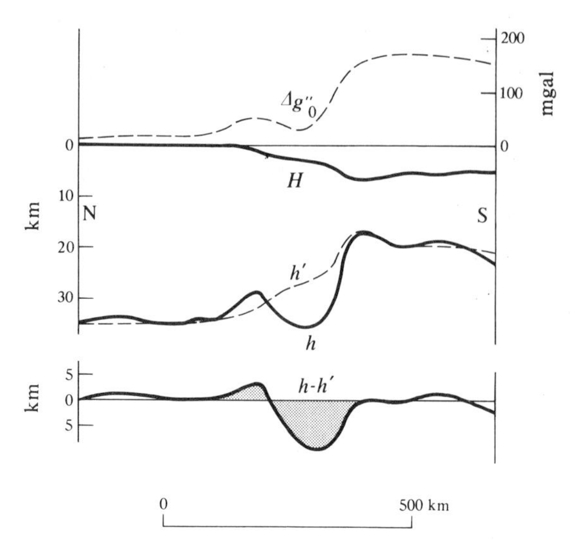

Figure 14.8 $\Delta g_0''$, H, h, h' and $(h - h')$ along Profile 17 of Vening Meinesz. h' represents the bottom surface of the Earth's crust assuming isostatic equilibrium, and h represents the bottom surface of the Earth's crust calculated from $\Delta g_0''$.

the Japan Trench (by M. Matuyama and N. Kumagai in 1933). A study of the crustal structure under the Japan Trench by S. Yamaguti in 1937 has been mentioned before.

New methods for gravity measurements on board surface ships have been developed so that the knowledge about the distribution of gravity anomalies over oceanic trenches is increasing greatly. Figure 14.9 shows the distribution of gravity anomalies over the Japan Trench and the neighbouring islands in detail (Tomoda 1973a,b). Tomoda has measured gravity values over several other trenches in the Western Pacific Ocean also and pointed out the following features.

(a) When an island arc and an oceanic trench are associated, Δg_0 is positive on the island side and is negative on the trench side. The areas where Δg_0 is larger than 50 mgal and where Δg_0 is less (that is more negative) than -50 mgal are nearly equal. This means that the excess mass under the island arc and the deficiency of mass under the trench are equal in amount and, in this sense, the mass is conserved within this combined system.

(b) The axis of largest negative Δg_0 and that of largest depth of the trench run parallel. In some cases, they coincide geographically, but in certain other cases, they do not coincide exactly. For example, in the trench which runs from the Idu Island Arc down to the Mariana Arc, these two axes coincide in position. But in the Japan Trench which, along its northern extension, runs along the Pacific

Figure 14.9 Gravity anomalies around the Japan Trench. $\Delta g_0''$ (mgal) on land, and Δg_0 (mgal) on the ocean. The shaded area represents a negative anomaly.

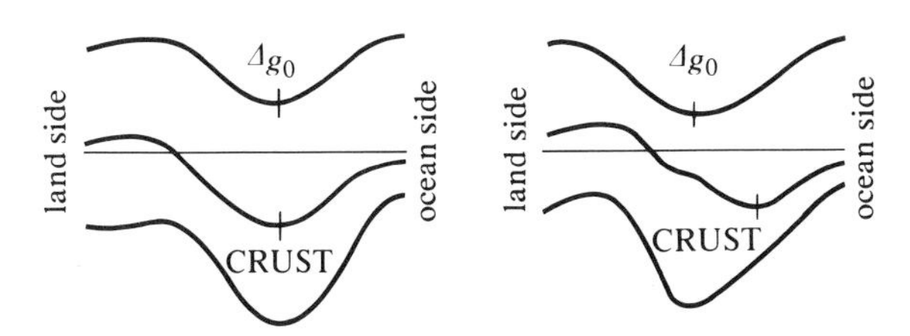

Figure 14.10 Relative position of a minimum value of Δg_0 and the trench axis.

side of the northern part of Honshu (Japan), they do not coincide and the axis of the most negative Δg_0 runs on the island side of the trench. From this point of view, trenches can be of two types, one is that which does not have land behind it and the other is that which does. Figure 14.10 illustrates these two types schematically.

The values of Δg_0 are not necessarily proportional to the depth of the trenches. According to Tomoda, this is an important fact concerning the developmental history of trenches in relation to their age of formation.

14.4 Gravity on sea-mounts

There are a number of isolated big mountains rising several thousand metres from the ocean bottom. These are called sea-mounts. As referred to the surrounding ocean bottom, sea-mounts are about 3000–4000 m high. The top surfaces of sea-mounts are about 1000 m deep and in some cases, there are flat plains wider than 10 km across. Y. Tomoda and J. Segawa made gravity measurements above several sea-mounts in the western Pacific in 1969. As an example, Figure 14.11 shows the results obtained above Suiko sea-mount, which is located 1000 km south-west of the southern end of Kamchatka Peninsula.

Above the sea-mount, Δg_0 is 300 mgal and this is due to the attraction of the mass of the sea-mount. A remarkable feature to notice is that there are areas in which Δg_0 is negative and this area extends far from the sea-mount. According to Tomoda, the mass of the sea-mount must have been squeezed out from these areas of negative gravity anomalies.

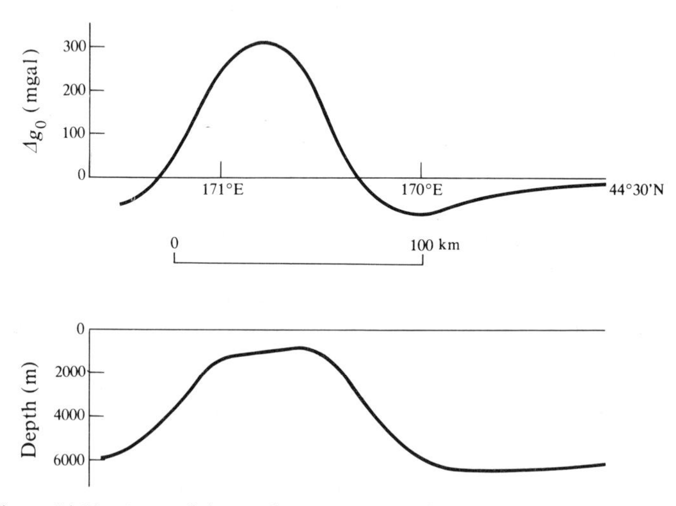

Figure 14.11 Δg_0 and the sea-bottom topography around Suiko sea-mount.

14.5 Gravity on marginal seas

A marginal sea is a sea lying between a continent and a bordering island arc, the Sea of Japan being an example. It is about 3000 m deep on average and is shallower than a standard ocean bottom. Δg_0 values on marginal seas are about 20–30 mgal.

The East China Sea, which lies between China and the Ryukyu Island Arc, is another example of marginal sea. It is only several hundred metres deep but the gravity distribution around it shows an oceanic characteristic. It is considered that this part is oceanic in geophysical nature but the tremendous amount of sediment with a density 2.0 coming from the Yangtze River, which have been deposited to a thickness of 2000 m as shown in Figure 14.12, make it look as if it were a shallow sea topographically.

In closing this chapter, it should be added that the problems concerning ocean-bottom structures are now being intensively investigated by a number of geophysicists throughout the world, especially in connection with the plate tectonic theory. It is very probable that what has been stated in this chapter, on the crust–mantle model, will have to be revised or supplemented in the not too-distant future.

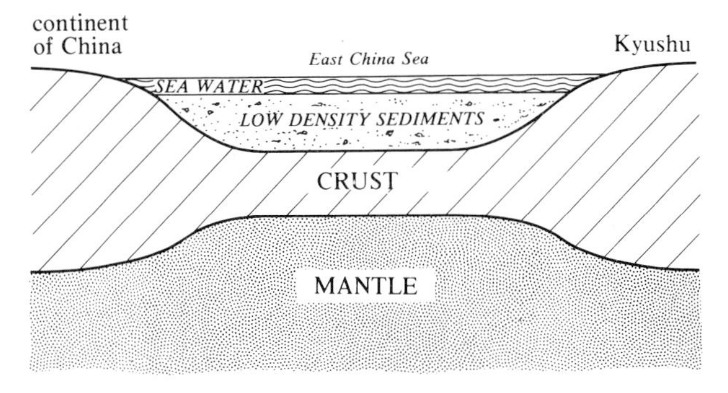

Figure 14.12 Cross section of the East China Sea.

References and further reading

Hess, H. H. 1933. *Interpretation of geological and geophysical observations*. Report Navy-Princeton Gravity Expedition to the West Indies in 1932, U.S. Navy Hydrographic Office, 27.

Hess, H. H. 1938. Gravity anomalies and island-arc structure with particular reference to the West Indies. *Proc. Am. Phil Soc.* **79**, 71.

Matuyama, M. 1936. Distribution of gravity over the Nippon Trench and related areas. *Proc. Imp. Acad., Japan* **12**, 93.

Talwani, M., X. Le Pichon and M. Ewing 1965. Crustal structure of the Mid-Oceanic Ridge. *J. Geophys. Res.* **70**, 341.

Tomoda, Y. 1973a. *Gravity anomalies in the Pacific Ocean, the Western Pacific*. Nedlands: University of Western Australia Press.

Tomoda, Y. 1973b. *Maps of free air and Bouguer anomalies in and around Japan*. Tokyo: University of Tokyo Press.

Tomoda, Y., J. Segawa and A. Tokuhiro 1970. Free air gravity anomalies at sea measured by the Tokyo surface ship gravity meter (1961–1969). *Proc. Jap. Acad.* **46**, 1006.

Tsuboi, C. 1939. Relation between the gravity anomalies and the corresponding subterranean mass distribution (III). *Bull. Earthquake Res. Inst.* **17**, 351.

Tsuboi, C. 1957. *Crustal structure along a certain profile across the East Indies as deduced by a new calculation method.* Gedenkboek F. A. Vening Meinesz, N. V. Boek-en Kunstdrkkerij, Mouton, 287.

Vening Meinesz, F. A. 1934. *Gravity expeditions at sea*, Vol. I. Neth. Geod. Comm.

Vening Meinesz, F. A. 1948. ibid., Vol. II.

15 Gravity in volcanic and earthquake areas

15.1 Volcanoes and gravity

Volcanoes are made up of an enormous amount of volcanic materials that have come out from the depths to the surface of the earth. The Hawaiian Islands, for example, are a group of volcanic bodies rising 5000 m from the bottom of the Pacific Ocean. The masses of which the islands are composed are well in excess of 10^{22} tonnes. The gravity values around the islands must show a characteristic distribution due not only to the attractions of the enormous masses but also to the migration of material from the depths to the surface. Similar characteristic distributions of gravity will be observed also around a number of volcanoes on land. Even after a volcano has been formed, eruptions occur there and occasionally enormous masses come out again from the depths or existing masses are blown off. Mass migrations are thus continually taking place around volcanoes.

Figure 15.1 shows the wide distribution of volcanic ejecta on Kyushu, Japan, from four volcanoes which are shown by dotted lines. The total volcanic mass ejected from one of them, volcano Aso, is estimated to be 3×10^{11} tonnes. This is the integrated value throughout geological time. Even one single eruption is sometimes associated with a considerable migration of mass. In the Sakurajima (Aira) eruption of 1914, lava and other volcanic materials ejected were estimated to weigh 5×10^{9} tonnes. Usu volcano, Hokkaido, Japan, on the other hand, has exhibited a peculiar feature in its activity. Instead of volcanic materials being ejected, a huge columnar mass of volcanic rock was pushed up to a height of several hundred metres.

Although there are many different types of volcanic eruption, they are common in that huge migrations of materials are involved. Gravity observations around volcanoes will give valuable information about how the migrations took place. Gravity measurements around volcanoes, however, have not been easy, because of the difficulties involved in transportation of heavy instruments for measuring gravity and the accuracy required in height determination of observation points in wild volcanic areas. Recently, the difficulties have been partly solved by the development of portable static gravimeters.

T. C. Mendenhall made a gravity measurement at the top of Mt Fuji in Japan in 1882 and from the result he obtained, he calculated the

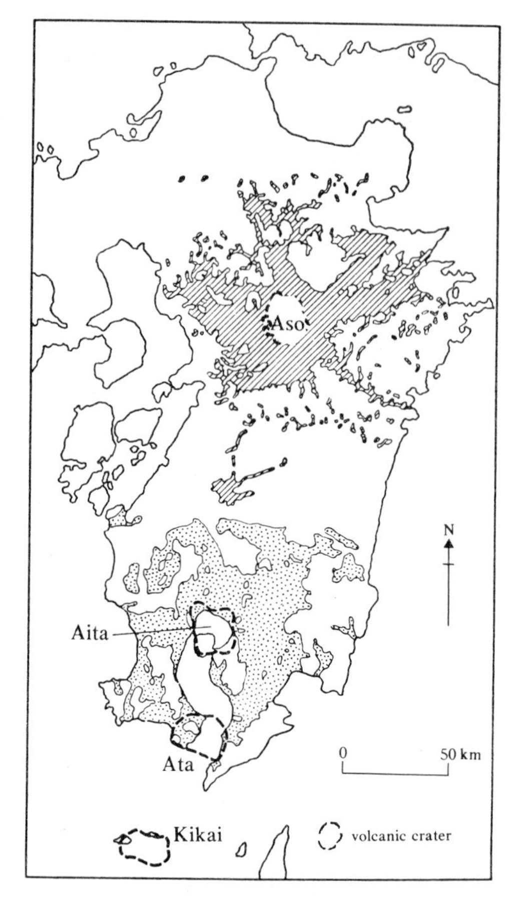

Figure 15.1 Distribution of volcanic ejecta on Kyusha, Japan.

mean density of the mountain to be $\rho = 2.08$ which is remarkably low compared with that of other non-volcanic mountains. Although there are some questionable points in his calculations, this measurement at Mt Fuji may be regarded as the first one in the study of gravity on volcanoes.

Forty years after this, I. Yamamoto made Eötvös torsion-balance surveys on the active volcano, Asama, in Japan in 1923. He repeated the measurements at several points on Asama and reported that the Eötvös quantities had changed by more than 200 Eötvös in association with the eruptions of the mountain. In 1936, T. Minakami made a more systematic torsion-balance survey at 26 points on Asama (Minakami 1937). His results are shown in Figure 11.14. It should be noted that the vectors representing the gradients of gravity are not directed toward the centre of the volcano. During the period of this measurement, a fairly large eruption took place on the volcano. Minakami reoccupied several stations where he had made observations before and found that the Eötvös quantities had changed by 2–10 Eötvös units, which must have taken place in association with the eruption.

Recently I. Yokoyama has made extensive pioneering studies related to gravity on volcanoes, and has visited about 40 volcanoes throughout the world. His results will be described in the subsequent sections.

15.2 Masses under volcanoes and Gauss's theorem

The distribution and magnitude of gravity anomalies around a volcano provide valuable data for investigating its geophysical structure. A problem of special interest related to volcanoes is whether there is a defect of mass or an excess of mass beneath a volcano, or whether there is no mass anomaly at all.

To solve this problem, Gauss's theorem is useful. As was mentioned before, this theorem states that if the gravity values on a surface which encloses a mass M are integrated over the whole surface, the result gives the value of $4\pi G$ times M. In the case of a volcano, Δg can be measured only over half of the surface so that M will be given approximately by

$$M = \frac{1}{2\pi G}\int_{-\infty}^{\infty}\int_{-\infty}^{\infty} \Delta g(x, y)\, \mathrm{d}x\, \mathrm{d}y \tag{15.1}$$

using 2π instead of 4π.

The merit of this method lies in the fact that the anomalous mass M itself can be found without knowing or assuming its shape and size.

15.3 Anomalous mass beneath volcanoes

In 1957, I. Yokoyama and H. Tajima made gravimetric measurements at 123 points on Mt Fuji. For making topographical corrections to the observed gravity values, they took the mean density of the mountain to be $\rho = 2.63$, instead of Mendenhall's value of 2.08. The new value of 2.63 was found from the variation of the gravity value according to the height on the mountain. The distribution of Bouguer anomalies $\Delta g_0''$ on this volcano is shown in Figure 15.2.

In this figure, the contour lines of $\Delta g_0''$ show little influence caused by the existence of the mountain itself, indicating that there is no noticeable mass anomaly beneath it. Yokoyama concluded that the volcanic materials of this mountain must have been supplied from a magma reservoir at the upper part of the mantle, through relatively narrow channels. According to Yokoyama, volcanoes Asama in Japan, Merbabu, Merapi in Central Java, and Etna in Italy show similar features. These volcanoes are mostly conical in shape and are often called strato-volcanoes.

There are other types of volcanoes which have broad basins called **calderas** (meaning pot in Spanish). The caldera is larger than 10 km across in many cases and its outer rim is in the form of a cliff surrounding the centre of the volcano. The distribution of $\Delta g_0''$ on the caldera volcano, Mihara in Ooshima, Japan, is as shown in Figure 15.3 (Yokoyama 1969). As a whole, $\Delta g_0''$ is positive and increases toward the east. This is the regional tendency in the distribution of $\Delta g_0''$ which is seen in the neighbourhood of the island. Superposed on this general distribution, a positive high of $\Delta g_0''$ is seen at the east rim of the caldera. The relative high is about 7 mgal. This gravity high can be accounted for by the attraction of a vertical cylindrical mass with a diameter 3 km and a density 0.3 higher than that of the surrounding rocks.

On Hawaii, $\Delta g_0''$ is as large as 200–300 mgal and this is due to the attraction of the enormous mass of the island. On the caldera of Kilauea,

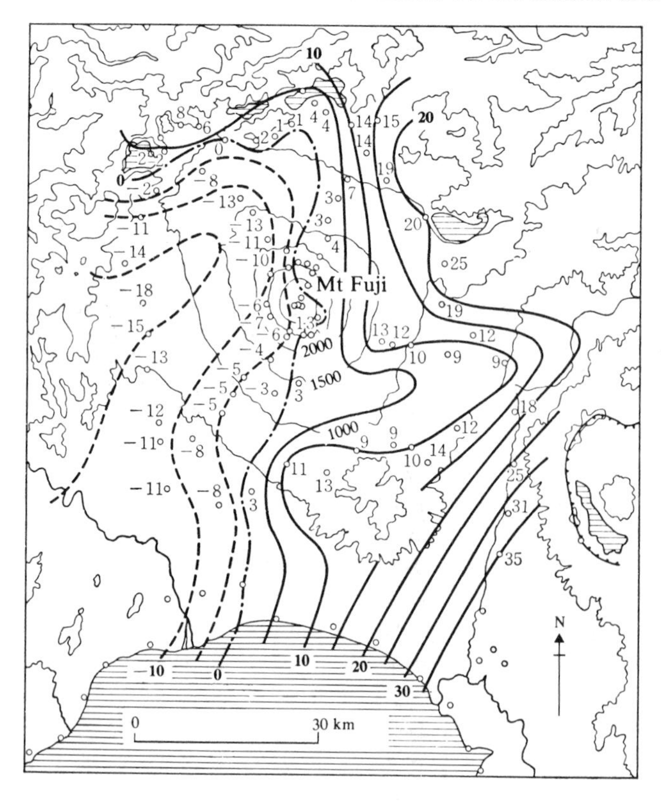

Figure 15.2 Distribution of Bouguer anomalies (in mgal) around Mt Fuji.

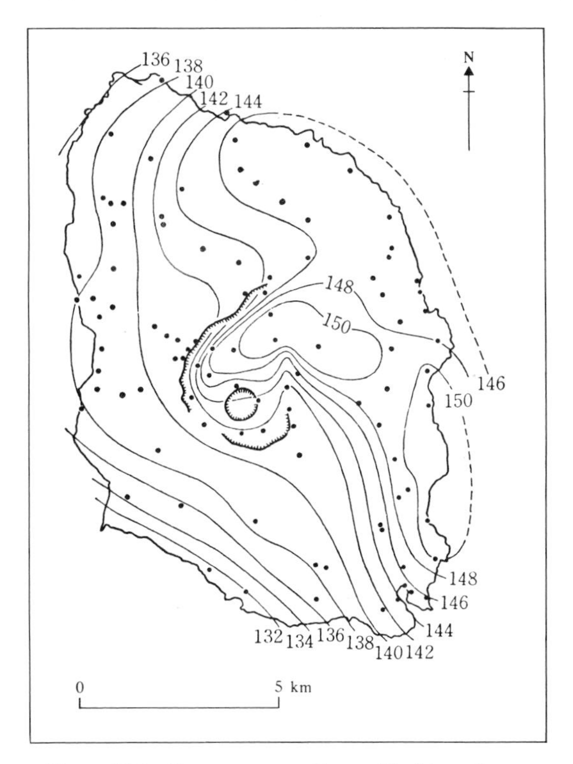

Figure 15.3 Bouguer anomalies on Ooshima, Japan.

$\Delta g_0''$ is about 10 mgal relatively higher than in the area surrounding. According to W. T. Kinoshita (1965), this relatively positive anomaly can be accounted for by hidden volcanic masses at a depth of 3 km with a thickness 2 km and a density 0.3 higher than the surrounding rocks. There are many other volcanoes in the world which have calderas and show relatively positive $\Delta g_0''$. $\Delta g_0''$ on volcanoes in Mull, Scotland, for example, amounts to 55 mgal.

There are also many volcanoes which have calderas and show a relatively negative $\Delta g_0''$. According to Yokoyama, the upper parts of these volcanoes were broken off by previous eruptions into a form of inverted funnel and loose volcanic ejecta of lower density filled these parts, thus causing a negative $\Delta g_0''$.

A good example of this is seen at Kuttcharo caldera, Hokkaido, Japan, which is about 22 km across and half of which is now a lake. Yokoyama made a gravity survey on the lake in the winter of 1958 when the lake water was frozen. Combining the lake survey with the land survey, the distribution of $\Delta g_0''$ found by Yokoyama (1969) is as shown in Figure 15.4. $\Delta g_0''$ at the centre of the caldera is as much as 45 mgal smaller than the surrounding area. This relative low can be

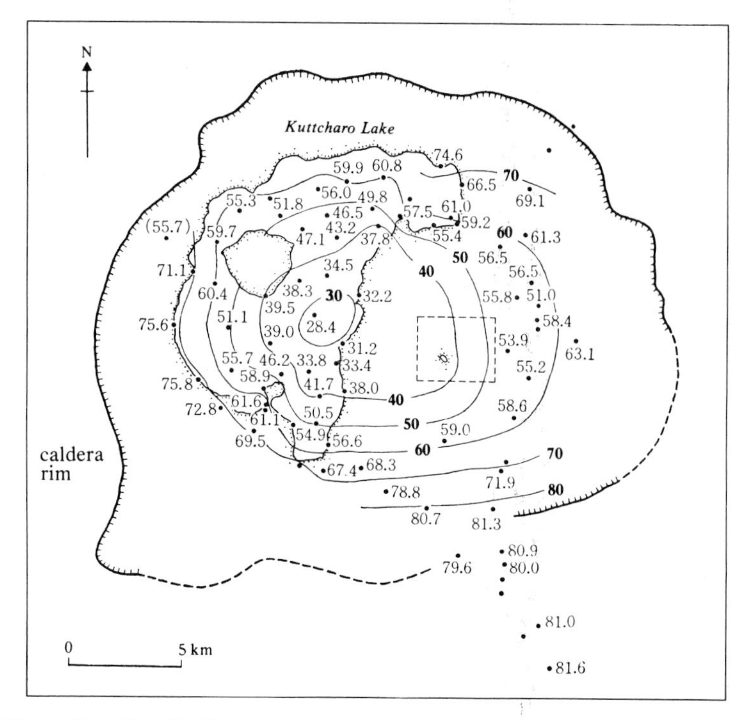

Figure 15.4 Distribution of Bouguer anomalies (in mgal) on Kuttcharo caldera.

accounted for by low-density material, with a density 0.4 lower than the surrounding rocks and a thickness 4 km at the centre, thinning out toward the rim of the caldera. The defect of the mass, calculated by Gauss's theorem, is of the order of 10^{11} tonnes.

On Aso volcano, which has the largest caldera in the world, A. Kubotera made an extensive gravity survey in 1969 at as many as 446 points. $\Delta g_0''$ shows a lowest value of -30 mgal at its centre.

Similar gravity lows have been found under many other volcanoes and they can be accounted for by loose volcanic materials deposited in calderas in the form of an inverted conical funnel.

Summarising these and other results, Yokoyama found that the negative mass M observed in ten volcanic calderas in the world can be related to its diameter D by the expression

$$M \propto D^{3.6}$$

as shown in Figure 15.5. It is noteworthy that the exponent 3.6 differs considerably from 3.0, indicating that the negative masses in calderas are not geometrically similar in shape.

There are topographical features which are called meteoric craters

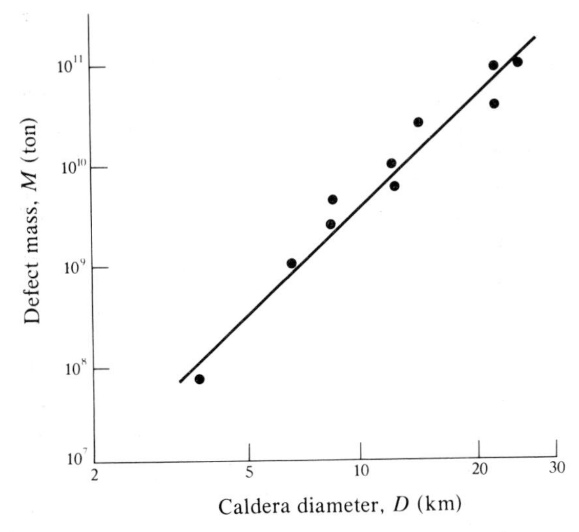

Figure 15.5 Relationship between the defect mass and the diameter of calderas.

and which are similar in geometrical shape to volcanic calderas but are much smaller in dimension. Gravity anomalies also are negative at meteoric craters, but in this case the negative mass at the craters and their diameters is related as follows:

$$M \propto D^{2.5}.$$

It should be noted that the exponent of D in this case is 2.5 and it differs from 3.6 in the case of volcanic calderas.

15.4 Volcanic eruptions and changes in gravity with time

Big volcanic eruptions are often associated with an outflow of lava and the ejection of broken fragments of volcanic materials. In the great eruption of Sakurajima volcano, Japan, in 1914, the enormous amount of lava which flowed out was estimated to be 5×10^9 tonnes. Such a large outflow of mass must have created changes in the distribution of gravity around the volcano.

There have been fragmental reports which state that gravity values around a volcano change in association with its eruption. This is a very important and interesting phenomenon but it must be remembered that the change can be caused apparently also by variation in height of the observation point before and after the eruption. A change in height by 10 cm could produce a gravity change of 30 μgal. Accurate determinations of both height and gravity must be made simultaneously before the cause of the change in gravity can be established.

15.5 Earthquake distribution and gravity

Figure 15.6 shows the zones in Japan across which gravity changes are spatially rapid. Figure 15.7 shows the distribution of earthquake epicentres, with $M > 5.0$, between 1900 and 1950. Figure 15.8 shows the vertical distribution of earthquake foci projected on a vertical plane along the Japanese islands.

Looking at these figures, it can be seen that rather shallow crustal earthquakes in the western half take place in the Earth's crust where the spatial changes in gravity are rapid, while in the northeastern part, earthquakes occur mostly in the mantle under those areas which are bounded by zones of steep horizontal gravity gradients.

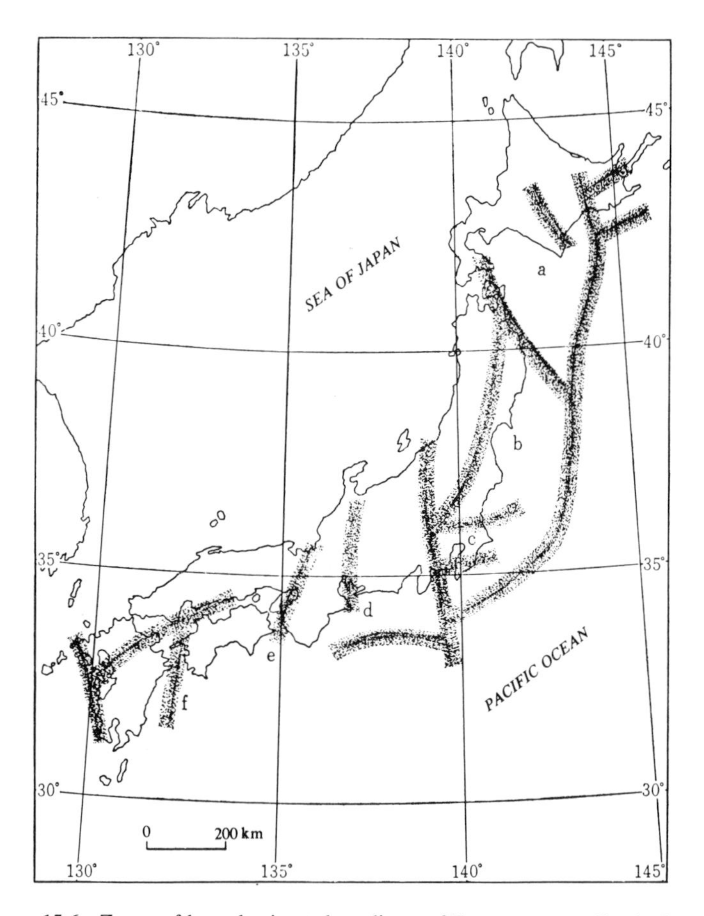

Figure 15.6 Zones of large horizontal gradients of Bouguer anomalies in Japan.

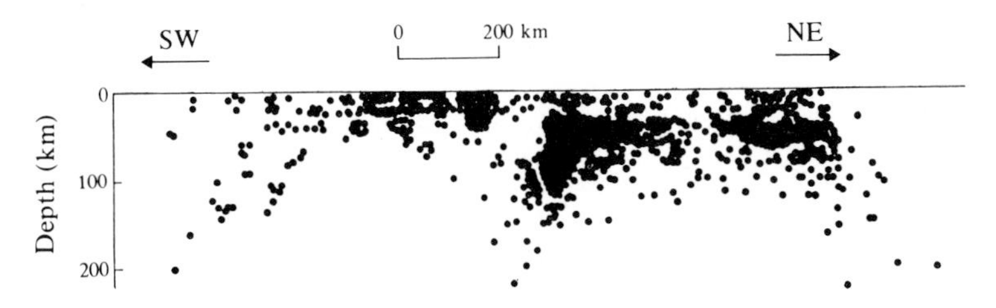

Figure 15.7 Distribution of epicentres in and around Japan, with a magnitude greater than 5 between 1900 and 1950.

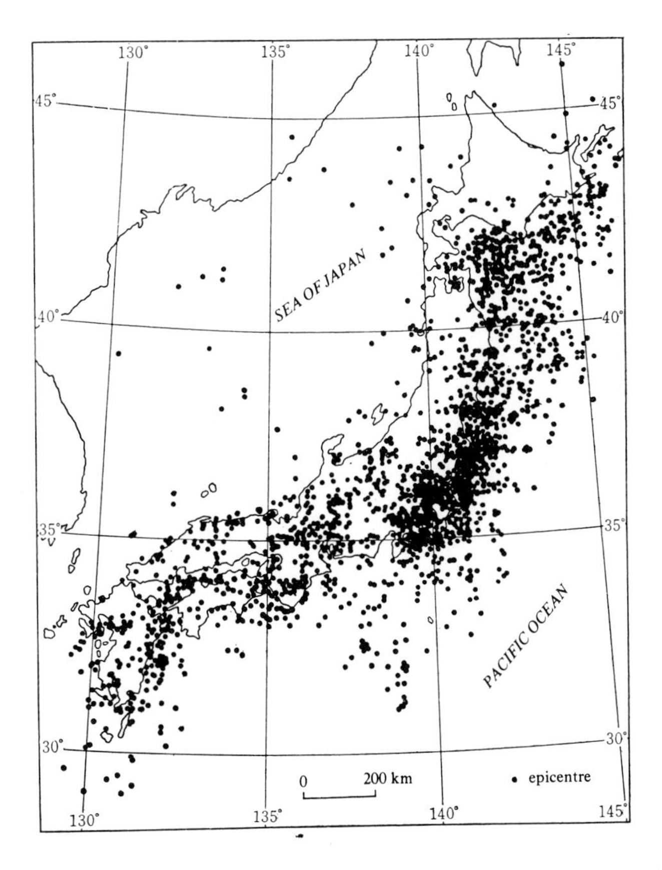

Figure 15.8 Distribution of earthquake foci projected onto a vertical plane along the Japanese islands.

References and further reading

Minakami, T. 1937. Torsion balance surveys on Volcano Asama. *Bull. Earthquake Res. Inst.* **15**, 50.
Tsuboi, C. 1929. On a relation between the distribution of gravitational anomalies and the origins of earthquakes in Japan. *Proc. Imp. Acad., Japan* **5**, 326.
Yokoyama, I. 1963a. Structure of caldera and gravity anomaly. *Bull. Volc.* **26**, 67.
Yokoyama, I. 1963b. Volcanic calderas and meteorite craters with the special relation to their gravity anomalies. *J. Fac. Sci., Hokkaido University, Series II* **2**, 37.
Yokoyama, I. 1966. Structure of calderas and their origin (in Japanese). *J. Volc. Soc., Japan* **10**, 119.
Yokoyama, I. 1974. Geomagnetic and gravity anomalies in volcanic areas. *Phys. Volc.* **15**, 41.

Bibliography

Cook, A. H. 1969. *Gravity and the Earth*. London: Wykeham.

Garland, G. D. 1965. *The Earth's shape and gravity*. London: Pergamon.

Heiskanen, W. and H. Moritz 1967. *Physical geodesy*. San Francisco: W. H. Freeman.

Heiskanen, W. and F. A. Vening Meinesz 1958. *The Earth and its gravity field*. New York: McGraw-Hill.

Helmert, F. A. 1884. *Die mathematischen und physikalischen Theoriien der höheren Geodäsie*. Leipzig: Teubner.

Jeffreys, H. 1952. *The Earth*, 3rd edn. Cambridge: Cambridge University Press.

National Research Council 1931. *The figure of the Earth*. Bull. U.S. National Research Council, no. 78. Washington, D.C.

Author index

Subject index